Extremophilic Microbes and Metabolites - Diversity, Bioprospecting and Biotechnological Applications

Edited by Afef Najjari, Ameur Cherif, Haïtham Sghaier and Hadda Imene Ouzari

Published in London, United Kingdom

IntechOpen

Supporting open minds since 2005

Extremophilic Microbes and Metabolites - Diversity, Bioprospecting and Biotechnological Applications
http://dx.doi.org/10.5772/intechopen.78716
Edited by Afef Najjari, Ameur Cherif, Haïtham Sghaier and Hadda Imene Ouzari

Contributors
Federico Buonanno, Claudio Ortenzi, Angeles Aguilera, Elena González-Toril, Graciela De Diego-Castilla, Susana Osuna, Suthyvann Sor Mendi, Rafael Bardera, Yolanda Blanco, Janmejay Pandey, Deepika Goyal, Shiv Swaroop, Afef Najjari, Hiba Mejri, Marwa Jabbari, Haitham Sghaier, Hadda Imene Ouzari, Ameur Cherif, Michelle J. Rivera-Rivera, Verónica Acosta-Martínez, Elvira Cuevas, Farah Faridi, Saba Khan, Sudipto Datta, Ranjit Barua, Jonali Das, Ahmed Mohammed Shuikan, Maram Sandouka, Sulaiman Al Harbi, Wael N. Nabil Hozzein, Sulaiman A. Al Yousef, Mohammed M. Alzharani, Eman Damra

First published in London, United Kingdom, 2021 by IntechOpen
IntechOpen is the global imprint of INTECHOPEN LIMITED, registered in England and Wales, registration number: 11086078, 5 Princes Gate Court, London, SW7 2QJ, United Kingdom
Printed in Croatia

British Library Cataloguing-in-Publication Data
A catalogue record for this book is available from the British Library

Additional hard and PDF copies can be obtained from orders@intechopen.com

Extremophilic Microbes and Metabolites - Diversity, Bioprospecting and Biotechnological Applications
Edited by Afef Najjari, Ameur Cherif, Haïtham Sghaier and Hadda Imene Ouzari
p. cm.
Print ISBN 978-1-83969-038-9
Online ISBN 978-1-83969-039-6
eBook (PDF) ISBN 978-1-83969-040-2

For EU product safety concerns:
IN TECH d.o.o., Prolaz Marije Krucifikse Kozulić 3, 51000 Rijeka, Croatia, info@intechopen.com or visit our website at intechopen.com.

Meet the editors

Dr. Afef Najjari is an Assistant Professor of Bioinformatics at the Faculty of Sciences of Tunisia, University of Tunis El Manar. Dr. Afef has worked on several topics including (i) genetic and enzymatic diversities of lactic acid bacteria isolated from dairy and meat products, and (ii) microbial diversity and ecology of extremophilic microbes in arid and saline ecosystems and mainly on archaeal groups. These works were funded by national and international projects. Currently, she is interested in metagenomic analysis, genome assemblies and annotations, transcriptomic data analysis (microarrays), and biological database development.

Dr. Ouzari Hadda-Imene has been a Professor of Microbiology at the Faculty of Sciences of Tunis, University Tunis El Manar since 2013. The main research and teaching activities of Prof. Ouzari Hadda-Imene are focused on the identification and biotechnological potentialities and applications of microbes from different environments. She has published about 90 peer-reviewed papers and more than 600 nucleotidic sequences. Prof. Ouzari Hadda-Imene has contributed to several national and international projects focusing on several aspects of microbial ecology. As an active member of the Tunisian Society of Microbial Ecology, she is actively contributing to the research promotion and scientific interaction between researchers from different countries. She also acts as the Tunisian ambassador of the International Society of Microbial Ecology (ISME).

Prof. Ameur Cherif graduated in Natural Sciences (1995) and has obtained his Ph.D. degree in Microbiology from the Faculty of Sciences of Tunis (FST), University of Tunis El Manar in 2001. He is the Head of a research laboratory "Biotechnology and Bio-Geo Resources Valorization" at the University of Manouba. His research activities focus on: (i) phylogeny and ecology of extremophilic microbes; (ii) microbial ecology of symbionts, commensal, and pathogen microbiota in different hosts; and (iii) biotechnological applications of active biomolecules, plant growth promotion, and bioremediation. He participated in the European Cost Action FA 0701, was a partner member in several EU FP7, Tempus, Erasmus, H2020 and bilateral projects, and serves on the various review and editorial committees. Prof. Cherif has published over 80 peer-reviewed articles and book chapters (h-index 21, Scopus, May 2018).

Dr. Ir. Haïtham Sghaier is an Engineer in Biotechnology and Professor (Associate) of Radiation and Computational Biology at the National Center for Nuclear Sciences and Technology (CNSTN) in Tunisia. Dr. SGHAIER has published on the evolution of ionizing-radiation-resistant Archaea and Bacteria. Also, he has published on bioremediation issues. Currently, he is implicated in various ongoing sequencing projects and databases [http://home.isima.fr/irrb/; http://www.radiop.org.tn/; etc.] about Archaea and Bacteria. In 2016, Dr. SGHAIER won the TWAS Young Arab Scientist (YAS) Prize.

Contents

Preface

Extremophilic microbes are microorganisms that can grow under a remarkable range of extreme environments such as glacial deserts, hot springs, ocean floors, hypersalted environments, and rocks of the Earth's mantle. These microbes include acidophiles, alkalophiles, halophiles, barophiles, (hyper), thermophiles, and psychrophiles. These hostile environments shelter the rich biodiversity of extremophiles that could belong to one of three domains of life (bacteria, archaea and eukaryote). These microbes have evolved several mechanisms to ensure genomic integrity, cell division, and energy conservation in extreme conditions. They present a wide and versatile metabolic and enzymatic diversity coupled with extraordinary physiological adaptabilities to extreme environments. These enzymes and metabolites have been exploited to develop clean and sustainable industrial processes. Antibiotics, compatible solutes, and other compounds obtainable from these microbes are also finding a variety of uses. Recently, several investigations have been started to study the phylogenetic relationship between extremophilic microbes through the analysis of their genome sequences. Hence, comparative genomic analyses of genomes allowed for the identification of distinctive genes and metabolic pathways involved in the extremophilic way of life. This book presents a detailed portrait of the diversity and metabolic potential and applications of extremophilic microbes through several approaches and helps the reader to understand the role of metabolites in environmental adaptations.

Afef Najjari and Hadda-Imene Ouzari
Department of Biology,
Faculty of Sciences of Tunis - University of Tunis El Manar,
Campus Universities,
Tunis, Tunisia

Haïtham Sghaier
National Center for Nuclear Sciences and Technology,
Sidi Thabet Higher Institute for Biotechnology,
Ariana, Tunisia

Ameur Cherif
Higher Institute for Biotechnology - University of Manouba,
Ariana, Tunisia

Chapter 1

Enhancement and Identification of Microbial Secondary Metabolites

Ahmed M. Shuikan, Wael N. Hozzein, Mohammed M. Alzharani, Maram N. Sandouka, Sulaiman A. Al Yousef, Sulaiman A. Alharbi and Eman Damra

Abstract

Screening for microbial secondary metabolites (SMs) has attracted the attention of the scientific community since 1940s. In fact, since the discovery of penicillin, intensive researches have been conducted worldwide in order to detect and identify novel microbial secondary metabolites. As a result, the discovery of novel SMs has been decreased significantly by using traditional experiments. Therefore, searching for new techniques to discover novel SMs was one of the most priority objectives. However, the development and advances of omics-based techniques such as metabolomics and genomics have revealed the potential of discovering novel SMs which were coded in the microorganisms' DNA but not expressed in the lab media or might be produced in undetectable amount by detecting the biosynthesis gene clusters (BGCs) that are associated with the biosynthesis of secondary metabolites. Nowadays, the development and integration of gene editing tools such as CRISPR-Cas9 in metabolomics provide a successful platform for the identification and detection of known and novel SMs and also to increase the production of SMs.

Keywords: secondary metabolites, metabolomics, genomic, CRISPR-Cas9, secondary metabolites identification, production of secondary metabolites, microorganisms, gene editing

1. Introduction

The term secondary metabolites (SMs) was first mentioned in 1891 by A. Kossel. Microbial secondary metabolites have attracted the scientific world's attention, since the discovery of penicillin in 1940s. After that, the identification and characterization of SMs have reached the highest level between 1940s and 1960s, and this period is called as "the golden era of SMs discovery" [1, 2]. A lot of compounds have been characterized and reported during the golden era and are still utilized till now. Unfortunately, the discovery of approved novel chemical scaffolds of secondary metabolites has significantly decreased after the golden era [1]. The possible explanation of the decreasing in the SMs' identification might be due to the following: (1) using the biosynthetic modules that are used for SMs' production

IntechOpen

by many bacteria, (2) focusing on some specific group of microorganisms such as actinobacteria, resulting in the isolation and identification of known compounds, and (3) almost 1% of the microbial community can be cultured in the laboratory [2] due to the difficulty in identifying their optimal medium compositions, resulting in the majority of SMs not being identified.

Microbial secondary metabolites play significant roles in our life. Because of SMs' unusual chemical structures, the microbial secondary metabolites show a variety of biological activities such as antimicrobial agents, antitumor agents, enzyme inhibitor, immunosuppressive agents, antiparasitic agents, herbicides, anthelmintic and food industry, etc. For instance, one of the huge successes in human medicine is the discovery of immunosuppression, such as cyclosporine A, which plays a significant role in establishing the organ transplant field.

All biochemical reactions carried out by organisms are called as metabolism and all products resulting from metabolism are called as metabolites. As a result of metabolism reactions, organisms produce primary and secondary metabolites. Primary metabolites are found in all living cells that are able to divide, while secondary metabolites are present only incidentally and do not affect the organism's life immediately. Nowadays, over 2,140,000 secondary metabolites (SMs) have been identified based on their vast diversity in function, structure, and biosynthesis (**Table 1**) [3]. The major sources of SMs are plants (about 80%) and microorganisms. Among microorganisms, bacteria, especially actinobacteria, and fungi have been reported to produce the majority of SMs that have been identified till now [4].

Microbial secondary metabolites (SMs) such as antibiotics, alkaloids, toxins, pigments, growth hormones, antitumor agents, and others are low molecular mass products that are produced by microorganisms, usually during the late growth phase. In fact, microbial secondary metabolites are not essential for the growth and development of microorganisms that produce them but are associated with some other functions such as competition, interactions, defense, and others [3, 5].

The development and advances of omics-based techniques such as genomics, metabolomics, proteomics, and trascriptomics have revealed that microorganisms have the potential to produce more secondary metabolites than were originally expected [6, 7]. These products are often coded by clustered genes present on the chromosomal DNA and rarely on plasmid DNA. In fact, most of these new SMs

Source	All known compounds	Bioactive
Plant kingdom	600,000–700,000	150,000–200,000
Microbes	Over 50,000	22,000–23,000
Higher plants	500,000–600,000	~100,000
Animal kingdom	300,000–400,000	50,000–100,000
Protozoa	Several hundreds	100–200
Vertebrates	200,000–250,000	50,000–70,000
Marine animals	20,000–25,000	7000–8000
Invertebrates	~100,000	NA
Algae, lichens	3000–5000	1500–2000
Insects, worms	8000–10,000	800–1000

NA—data not available.
Source: Bérdy [4].

Table 1.
Approximate number of identified natural metabolites.

have been predicted only by using bioinformatics analysis, which analyzes the putative SMs gene clusters in a sequenced genome. This is because, all of the new revealed SMs are not produced naturally under the lab conditions, or even though they are produced, this in very low amount that the traditional detection techniques are unable to detect them [8, 9]. Metabolomics approach aims to discover and characterize secondary metabolites in natural or engineered biosystems, and it can measure as many low molecular weight compounds as possible. Metabolomics-based technologies such as mass spectrometry (MS) and nuclear magnetic resonance (NMR) have been identified as significant analytical methods to detect SMs produced under specific conditions [10]. The present chapter provides an overview of present-day metabolomic and genetic engineering approaches for secondary metabolites' enhancement and identification [11].

2. Genomic for screening and enhancement of SMs

2.1 Gene editing for metabolites discovery

Biosynthesis gene clusters (BGCs) are the genes associated with the biosynthesis of secondary metabolites. These BGCs include all genetic information necessary for SMs' biosynthesis, assembly, modification, and regulation of their export and transport [11]. Microorganisms' genome contains variety of cryptic or silent genes that are responsible for the production of secondary metabolites but are not expressed under laboratory conditions. It has been reported that most BGCs remain silent and cannot be fully expressed under standard laboratory conditions. These silent BGCs are potentially significant in the discovery of novel SMs [11–14].

Due to the development of genomic and bioinformatic field, we are able to access extensive sequencing data and genetic information and enable genome mining of relevant BGCs with the potential for valuable SM production [15]. Therefore, biosynthetic biology and genetic engineering tools are now utilized for identification of novel BGCs. In fact, genetic engineering is now widely used and moving beyond traditional tools, which has opened a new era in the detection of novel secondary metabolites [16]. Genetic engineering for the production of SMs can be carried out in heterologous as well as homologous host. In fact, gene manipulation in heterologous host enables the activation of biosynthesis gene clusters (BGCs) obtained from unculturable organisms, whereas gene manipulation in homologous host allows the retention of all natural factors essential for the production of secondary metabolites [17]. While there is no single approach that will work for all genes of interest, a variety of techniques have been developed to induce the expression of these genes.

In fact, several genome techniques have emerged and are utilized in the metabolomic production field, including transcriptional activator-like effector nucleases (TALENs), zinc finger nucleases (ZFNs), and clustered regulatory interspaced short palindromic repeat (CRISPR-Cas9) [18, 19]. Each genome engineering technology has its own advantages and disadvantages (**Table 2**). For instance, ZFNs and TALENs have been successfully utilized in various microbes but still have limitation which includes the difficulty to engineer them [20]. Recently, CRISPR-Cas9 has been reported to be a significant and promising genome editing technology in the discovery and production of SMs [17, 16, 21].

2.1.1 Gene cloning

Direct cloning of the entire BGCs into the heterologous host is the most general and widely used approach for the activation of silent BGCs. Nowadays, many

		CRISPR/Cas9	Zinc finger nucleases (ZFNs)	Transcription factor-like effector nucleases (TALENs)
1	Protein engineering steps	It does not require protein engineering steps, very simple to test multiple gRNA	It requires complex steps to test gRNA	TALENs need protein engineering steps to test gRNA
2	Mode of action	It works by including double-strand breaks in target DNA or single-strand DNA nicks (Cas9 nickase)	It can induce double-strand breaks in target DNA	Induces DSBs in target DNA
3	Cloning	Not Required	Required	Required
4	Structural proteins	CRISPR consists of single monomeric protein and chimeric RNA	ZFNs work as dimeric and only protein component is required	TALENs also work as dimeric and require protein component
5	Mutation rate	Low mutation rate has been observed	High mutation rate has been observed in plants	Mutation rate is high as compared to CRISPR
6	Components	crRNA, Cas9 proteins	Zn-finger domains, nonspecific FOKI nuclease domain	Zn-finger domains, nonspecific folk nuclease domain
7	Length of target sequence (bp)	20–22	18–24	24–59
8	Target recognition efficiency	High	High	High
9	Level of experiment	Easy and very fast procedure	Complicated procedure and need for expertise in protein engineering	Relatively easy procedure
10	Methylated DNA cleavage	It can cleave methylated DNA in human cells. This aspect is of special concern for plants as this has not been much explored	Unable to do so	There are many question marks upon the capacity of TALENs to perform methylated DNA cleavage
11	Multiplexing	This is the main advantage of CRISPR, and several genes can be edited at same time. Only Cas9 is needed	Highly difficult to achieve this through ZFNs	Very difficult to obtain multiplexed genes by means of TALENs. Because it needs separate dimeric proteins specific for each target

Table 2.
Different genomic engineering techniques used in Metabolomic.

new cloning tools have been introduced, including Cas9-assisted targeting of chromosome segments (CATCH), transformation-assisted recombination (TAR), and TAR-CRISPR [22–24]. Basically, gene cloning steps include: determining the suitable heterologous host, cloning of the target BGC, transfer of the BGC into the chosen host, expression in chosen host system, and optimization of production.

Cas9-assisted targeting of chromosome segments (CATCH) is a cloning technique that utilizes the CRISPR-Cas9 gene editing system for direct BGCs cloning into the host. Comparing with traditional cloning tools such as PCR and restricted enzymes, CATCH is predicted to become a useful molecular tool for direct cloning of large gene clusters. Transformation-assisted recombination (TAR) has been used for cloning of large BGCs for about decades. However, the TAR approach is associated with a low cloning efficiency, which means it requires screening of hundreds of colonies to detect few positive clones [22, 24]. To address this challenge, TAR and CRISPR-Cas9 have been coupled resulting in a new approach called TAR-CRISPR [24]. By coupling TAR with CRISPR, a significant increase of the clone efficiency has been reported. Comparing with traditional TAR cloning, the advantages of TAR-CRISPR are that the positive clones could be achieved with secondary screening and lesser manpower and also it does not require a high experience of working with yeast [24]. In fact, the TAR-CRISTAR cloning will allow for the development of BGC cloning and SM production in the future.

2.1.2 Gene refactoring

Gene refactoring or replacement is useful not only in BGCs' activation but also for novel SMs' discovery. In fact, several silent BGCs have been refactored by replacing the BGC promoter to yield natural products such as secondary metabolites [25–28].

Another new tool in gene refactoring is multiplexed CRISPR-Cas9- and transformation-associated recombination (TAR)-mediated promoter engineering method (mCRISTAR) [21]. This new tool combines the advantages of the CRISPR-Cas9 system and TAR. It is different than the TAR-CRISPR that was discussed earlier. Comparing with TAR-CRISPR, which is a yeast-based method, basically mCRISTAR uses CRISPR-Cas9 to break the double-stranded in the promoter region of the BGC, and the fragments produced are reassembled by TAR with synthetic gene-cluster-specific promoter cassettes. Another gene refactoring tool that has aided in the faster cloning and refactoring of BGCs is the direct pathway cloning (DiPaC). Direct pathway cloning (DiPaC) depends on PCR amplification and in vitro DNA assembly for biosynthesis gene cluster capture and their expression. DiPaC was recently employed for the capture of biosynthesis gene cluster, which is small in size, followed by their activation and expression of novel natural products [29]. DiPaC was also able to successfully clone mid and large size of BGC [29].

2.1.3 Gene insertion or deletion

A large number of researches have documented the effect of gene knockout/in on BGC expression or levels of SM production. However, conventional methods of gene editing are time-intensive, while CRISPR-Cas9-based approach allows for much faster and efficient gene editing [30]. The emergence of CRISPR-Cas9 has opened up a new era in gene editing opportunities [31]. Recently, CRISPR gene editing approach has been used to insert promoter in order to activate microorganisms' SMs' production [32].

Nowadays, CRISPR-Cas9 is used to introduce promoter at multiple BGCs, and at the same time, resulting in the activation of BGCs followed by the production of SMs [32]. Multiplexed site-specific genome engineering (MSGE) was also used for multiple BGCs' editing [33]. MSGE has led to a significant increase in the secondary metabolites' production.

While, gene editing approaches provide a significant platform to manipulate the genetic machinery of microbes toward the production of novel, natural secondary metabolites, the identification of secondary metabolites is also equally important. Metabolomics plays a significant role in the identification and characterization of secondary metabolites produced by native or genetically modified microorganisms.

2.2 Identification and characterization of secondary metabolites

Unlike all omics techniques, metabolomics often requires a broad array of instrumentation such as coulometric array detectors for detecting redox compounds, fluorescent spectrometers for detecting aromatic compounds, and ELSD for detecting lipids, whereas genomics, proteomics, or transcriptomics measurements are often conducted by a single instrument.

In general, microbial secondary metabolites' investigation is mainly conducted in two different approaches, the targeted and untargeted metabolites' identification [34]. Targeted metabolites' experiments aim to detect a specific group of compounds (about 20 compounds) that are already identified. Whereas, untargeted secondary metabolites' investigation aims to detect and identify a large scale of metabolites that are produced by microorganisms, including known and novel metabolites [35].

Over the past decade, two general technologies have emerged as the primary tools in metabolomics, the nuclear magnetic resonance (NMR) spectroscopy and mass spectrometry (MS) [36]. Some of the common MS-based analyses are (GC-MS), (CE-MS), and (LC-MS) [37, 38]. These high-throughput tools provide a broad coverage of many classes of secondary metabolites, including amino acids, lipids, sugars, organic acids, and others.

2.2.1 Detection of secondary metabolites

Mass spectrometry (MS) is a technique that measures the mass-to-charge ratio of molecule. The principle of chromatography is to detect the retention time of the constituents that travel at different speeds under a specific condition. Therefore, various constituents take different time to pass from the inlet to the detector of the chromatography system [36].

Nuclear magnetic resonance (NMR) spectroscopy principle is based on using the magnetic properties of atomic nuclei to determine the chemical and physical properties of atoms or molecules in which they are contained. NMR's mechanism of action is that the magnetic nuclei in magnetic field absorb, resulting in reemitting of electromagnetic radiation at a specific resonance frequency depending on the magnetic properties of the isotope of atom as well as the strength of the magnetic field [36].

Both MS and NMR can be utilized to identify targeted and untargeted metabolomics. In fact, MS and NMR are often complementary techniques to each other. While NMR can be used to differentiate between structural isomers, MS provides information on the formula of the molecule [39]. Comparing to NMR, mass spectrometry is more sensitive and is able to detect a large scale of metabolites. On the other hand, nuclear magnetic resonance (NMR) spectroscopy is highly quantitative and reproducible. Unlike MS, NMR requires a larger sample amount for analysis [40, 41].

2.2.2 Data analysis

In fact, the complexity and huge amount of information that are obtained from either NMR spectroscopy or MS are considered to be one of the major challenges in metabolomics experiments [7]. The extraction of the important information that is generated by MS or NMR spectroscopy depends on using computer software in order to organize the vast amount of data [40]. First, the row data acquired from the NMR spectroscopy or MS must be first converted into computer formats compatible with software packages. In fact, the goal of metabolomics data analysis is to compare and identify the differences between hundreds or thousands of SMs. It is unpractical to visualize changes between groups of metabolites by analyzing metabolites individually; therefore, univariate and multivariate statistical techniques can then be used to interpret the data. One of the most widely used statistical methods is the principal component analysis (PCA) [39, 42, 43]. By using PCA, the data can be simplified without losing their main features. Generally, the PCA principle is based on reducing the dimensionality of the data set, while keeping characteristics participating most to the variance. In fact, PCA provides information on multivariate differences among metabolites. It is usually conducted at the early stages of data analysis.

However, different univariant statistical tests can be used to analyze isolated metabolites such as ANOVA, nonparametric Wilcoxon signed-rank test, Kruskal-Wallis test, and the parametric Student's t-test [44]. Furthermore, other univariant analysis can be used to validate the analysis such as false discovery rate calculations or Bonferroni correction [44].

2.2.3 Metabolites' identification

Due to the development of various bioinformatics software, most of metabolites can be identified. Two types of metabolites' identification are applied, including (a) putative identification and (b) definitive identification [7]. In putative identification, one or two molecular properties are utilized for identification. However, in definitive identification, two properties such as the retention time and accurate mass and/or fragmentation mass spectrum and/or NMR spectrum are used and compared with authentic chemical standard. Comparing to putative identification, definitive identification is a more accurate form of identification, while definitive identification uses the authentic chemical standard. Usually, the definitive identification is performed after the putative identification.

Nowadays, a variety of different metabolomics' databases are available online [45, 46].

Some are spectral-based databases as well as chemical structure-based databases for metabolites' identification. Generally, spectra generated during analysis are compared with reference compounds in databases, and then similarity is assigned to each other. Even though, metabolome databases are updated daily, still significant numbers of secondary metabolites in biological system are unidentified.

Some of the common databases used in nuclear magnetic resonance (NMR) spectroscopy are METLIN (http://metlin.scripps.edu), the Human Metabolome Database (HMDB, http://www.hmdb.ca), and Biological Magnetic Resonance Databank (http://www.bmrb.wisc.edu/metabolomics/), whereas commonly used databases for mass spectrometry are NIST (http://www.nist.gov/srd/nist1a.htm), MassBank (http://www.massbank.jp), the Golm Metabolite Database (GMD, http://csbdb.mpimp-golm.mpg.de/csbdb/gmd/gmd.html), METLI, and MMCD (http://mmcd.nmrfam.wisc.edu) [45].

3. Conclusion

Microorganisms are a rich source of secondary metabolites which have significant pharmaceutical, biomedical, and food applications. Nowadays, the development and integration of gene editing tools, especially CRISPR-Cas9 (gene cloning, gene refactoring, and gene insertion or deletion) in metabolomics, provide a successful platform for the identification and detection of known and novel SMs and also to increase the production of SMs. However, there are still some challenges associated with the application of metabolomics and gene editing, including that complete identification of novel SMs requires a combination of different methods which also result in increase in the screening cost. Thus, a comprehensive and sensitive technique is the need of hour, which has the ability to provide comprehensive information of any SMs under any conditions. Also, the off-target effect of CRISPR-Cas9 is a significant problem. However, the integration of metabolomics and CRISPR-Cas9-based gene editing tools may improve the efficiency of microbial secondary metabolites' discovery.

Author details

Ahmed M. Shuikan[1*], Wael N. Hozzein[2,3], Mohammed M. Alzharani[4], Maram N. Sandouka[5], Sulaiman A. Al Yousef[6], Sulaiman A. Alharbi[1] and Eman Damra[1]

1 Botany and Microbiology Department, College of Science, King Saud University, Riyadh, Saudi Arabia

2 Bioproducts Research Chair, Zoology Department, College of Science, King Saud University, Riyadh, Saudi Arabia

3 Botany and Microbiology Department, Faculty of Science, Beni-Suef University, Beni Suef, Egypt

4 Biology Department, College of Science, Al Imam Mohammad Ibn Saud University, Riyadh, Saudi Arabia

5 Department of Chemistry, Université du Québec à Montréal, Montréal, Quebec, Canada

6 Department of Clinical Laboratory Sciences, College Applied Medical Science, University of Hafr Albatin, Hafar Al-batin, Saudi Arabia

*Address all correspondence to: ahmedshuikan@gmail.com

References

[1] Li JWH, Vederas JC. Drug discovery and natural products: End of an era or an endless frontier? Biomeditsinskaya Khimiya. 2011;**57**(2):148-160

[2] Pelaez F. The historical delivery of antibiotics from microbial natural products - Can history repeat? Biochemical Pharmacology. 2006;**71**(7):981-990

[3] McMurry JE. Organic chemistry with biological applications. In: Secondary Metabolites: An Introduction to Natural Products Chemistry. Stamford, USA: Cengage Learning Ltd; 2015. pp. 1016-1046

[4] Bérdy J. Bioactive microbial metabolites. The Journal of Antibiotics. 2005;**58**(1):1-26

[5] Berg M, Vanaerschot M, Jankevics A, Cuypers B, Breitling R, Dujardin J-C. LC-MS metabolomics from study design to data-analysis – Using a versatile pathogen as a test case. Computational and Structural Biotechnology Journal. 2013;**4**:e201301002

[6] Putri SP, Nakayama Y, Matsuda F, Uchikata T, Kobayashi S, Matsubara A, et al. Current metabolomics: Practical applications. Journal of Bioscience and Bioengineering. 2013;**115**:579-589

[7] Go EP. Database resources in metabolomics: An overview. Journal of Neuroimmune Pharmacology. 2010;**5**(1):18-30

[8] Lim FY, Sanchez JF, Wang CCC, Keller NP. Toward awakening cryptic secondary metabolite gene clusters in filamentous fungi. Methods in Enzymology. 2012;**517**:303-324

[9] Gottelt M, Kol S, Gomez-Escribano JP, Bibb M, Takano E. Deletion of a regulatory gene within the cpk gene cluster reveals novel antibacterial activity in *Streptomyces coelicolor* A3(2). Microbiology. 2010;**156**:2343-2353

[10] Lenders J, Frédérich M, De Tullio P. Nuclear magnetic resonance: A key metabolomics platform in the drug discovery process. Drug Discovery Today: Technologies. 2015;**13**:39-46

[11] Bino RJ, Hall RD, Fiehn O, Kopka J, Saito K, Draper J, et al. Potential of metabolomics as a functional genomics tool. Trends in Plant Science. 2004;**9**(9):418-425

[12] Tran PN, Yen MR, Chiang CY, Lin HC, Chen PY. Detecting and prioritizing biosynthetic gene clusters for bioactive compounds in bacteria and fungi. Applied Microbiology and Biotechnology. 2019;**103**:3277-3287

[13] Valayil JM. Activation of microbial silent gene clusters: Genomics driven drug discovery approaches. Journal of Biochemistry & Analytical Biochemistry. 2016;**5**:276

[14] Rutledge PJ, Challis GL. Discovery of microbial natural products by activation of silent biosynthetic gene clusters. Nature Reviews. Microbiology. 2015;**13**:509-523

[15] Blin K, Kim HU, Medema MH, Weber T. Recent development of antiSMASH and other computational approaches to mine secondary metabolite biosynthetic gene clusters. Briefings in Bioinformatics. 2017:1103-1113. DOI: 10.1093/bib/bbx146

[16] Tong Y, Weber T, Lee SY. CRISPR/Cas-based genome engineering in natural product discovery. Natural Product Reports. 2019;**36**:1262-1280

[17] Zhang MM, Wang Y, Ang EL, Zhao H. Engineering microbial hosts for production of bacterial natural

products. Natural Product Reports. 2016;**33**:963-987

[18] Miller JC, Holmes MC, Wang J, Guschin DY, Lee YL, Rupniewski I, et al. An improved zinc-finger nuclease architecture for highly specific genome editing. Nature Biotechnology. 2007;**25**:778-785

[19] Jankele R, Svoboda P. TAL effectors: Tools for DNA targeting. Briefings in Functional Genomics. 2014;**13**:409-419

[20] Voytas DF. Plant genome engineering with sequence-specific nucleases. Annual Review of Plant Biology. 2013;**64**:327-350

[21] Kang HS, Charlop-Powers Z, Brady SF. Multiplexed CRISPR/Cas9-and TAR-mediated promoter engineering of natural product biosynthetic gene clusters in yeast. ACS Synthetic Biology. 2016;**5**:1002-1010

[22] Yamanaka K, Reynolds KA, Kersten RD, Ryan KS, Gonzalez DJ, Nizet V, et al. Direct cloning and refactoring of a silent lipopeptide biosynthetic gene cluster yields the antibiotic taromycin A. Proceedings of the National Academy of Sciences of the United States of America. 2014;**111**:1957-1962

[23] Jiang W, Zhao X, Gabrieli T, Lou C, Ebenstein Y, Zhu TF. Cas9-assisted targeting of chromosome segments CATCH enables one step targeted cloning of large gene clusters. Nature Communications. 2015;**6**:810

[24] Bonet B, Teufel R, Crüsemann M, Ziemert N, Moore BS. Direct capture and heterologous expression of *Salinispora* natural product genes for the biosynthesis of enterocin. Journal of Natural Products. 2015;**78**:539-542

[25] Lee NC, Larionov V, Kouprina N. Highly efficient CRISPR/Cas9-mediated TAR cloning of genes and chromosomal loci from complex genomes in yeast. Nucleic Acids Research. 2015;**43**:e55-e55

[26] Horbal L, Marques F, Nadmid S, Mendes MV, Luzhetskyy A. Secondary metabolites overproduction through transcriptional gene cluster refactoring. Metabolic Engineering. 2018;**49**:299-315

[27] Shao Z, Rao G, Li C, Abil Z, Luo Y, Zhao H. Refactoring the silent spectinabilin gene cluster using a plug-and-play scaffold. ACS Synthetic Biology. 2013;**2**:662-669

[28] Bauman KD, Li J, Murata K, Mantovani SM, Dahesh S, Nizet V, et al. Refactoring the cryptic streptophenazine biosynthetic gene cluster unites phenazine, polyketide, and nonribosomal peptide biochemistry. Cell Chemical Biology. 2019;**26**:724-736

[29] Greunke C, Duell ER, D'Agostino PM, Glöckle A, Lamm K, Gulder TAM. Direct pathway cloning (DiPaC) to unlock natural product biosynthetic potential. Metabolic Engineering. 2018;**47**:334-345

[30] Pohl C, Kiel JAKW, Driessen AJM, Bovenberg RAL, Nygard Y. CRISPR/Cas9 based genome editing of *Penicillium chrysogenum*. ACS Synthetic Biology. 2016;**5**:754-764

[31] Sander JD, Joung JK. CRISPR-Cas systems for editing, regulating and targeting genomes. Nature Biotechnology. 2014;**32**:347-355

[32] Zhang MM, Wong FT, Wang Y, Luo S, Lim YH, Heng E, et al. CRISPR–Cas9 strategy for activation of silent Streptomyces biosynthetic gene clusters. Nature Chemical Biology. 2017;**13**:607

[33] Li L, Zheng G, Chen J, Ge M, Jiang W, Lu Y. Multiplexed sitespecific genome engineering for overproducing bioactive secondary metabolites in

actinomycetes. Metabolic Engineering. 2017;**40**:80-92

[34] Breitling R, Ceniceros A, Jankevics A, Takano E. Metabolomics for secondary metabolite research. Metabolites. 2013;**3**:1076-1083

[35] Wu C, Kim HK, van Wezel GP, Choi YH. Metabolomics in the natural products field—A gateway to novel antibiotics. Drug Discovery Today: Technologies. 2015;**13**:11-17

[36] Dunn WB, Broadhurst DI, Atherton HJ, Goodacre R, Griffin JL. Systems level studies of mammalian metabolomes: The roles of mass spectrometry and nuclear magnetic resonance spectroscopy. Chemical Society Reviews. 2011;**40**(1):387-426

[37] Lu W, Bennett BD, Rabinowitz JD. Analytical strategies for LC-MS-based targeted metabolomics. Journal of Chromatography. B, Analytical Technologies in the Biomedical and Life Sciences. 2008;**871**(2):236-242

[38] Dettmer K, Aronov PA, Hammock BD. Mass spectrometry based metabolomics. Mass Spectrometry Reviews. 2007;**26**(1):51-78

[39] Alia A, Ganapathy S, de Groot HJ. Magic angle spinning (MAS) NMR: A new tool to study the spatial and electronic structure of photosynthetic complexes. Photosynthesis Research. 2009;**102**(2-3):415-425

[40] Dunn WB, Broadhurst D, Begley P, Zelena E, Francis-McIntyre S, Anderson N, et al. Procedures for large-scale metabolic profiling of serum and plasma using gas chromatography and liquid chromatography coupled to mass spectrometry. Nature Protocols. 2011;**6**(7):1060-1083

[41] Johnson CH, Gonzalez FJ. Challenges and opportunities of metabolomics. Journal of Cellular Physiology. 2012;**227**(8):2975-2981

[42] Hotelling H. Analysis of a complex of statistical variables into principal components. Journal of Education & Psychology. 1933;**24**:417-441

[43] Young SP, Wallace GR. Metabolomic analysis of human disease and its application to the eye. Journal of Ocular Biology, Diseases, and Informatics. 2009;**2**(4):235-242

[44] Broadhurst DI, Kell DB. Statistical strategies for avoiding false discoveries in metabolomics and related experiments. Metabolomics. 2006;**2**(4):171-196

[45] Brown M, Dunn WB, Dobson P, Patel Y, Winder CL, Francis-McIntyre S, et al. Mass spectrometry tools and metabolite-specific databases for molecular identification in metabolomics. Analyst. 2009;**134**(7):1322-1332

[46] Midelfart A, Dybdahl A, Gribbestad IS. Metabolic analysis of the rabbit cornea by proton nuclear magnetic resonance spectroscopy. Ophthalmic Research. 1996;**28**(5): 319-329

Chapter 2

Predator-Prey Interactions in Ciliated Protists

Federico Buonanno and Claudio Ortenzi

Abstract

Protists appeared relatively early in evolution, about 1.8 billion years ago, soon after the first prokaryotic organisms. During this time period, most species developed a variety of behavioral, morphological, and physiological strategies intended to improve the ability to capture prey or to avoid predation. In this scenario, a key role was played by specialized ejectable membrane-bound organelles called extrusomes, which are capable of discharging their content to the outside of the cell in response to various stimuli. The aim of this chapter is to describe the two main strategies adopted in ciliate predator-prey interactions: (a) the first is mediated by mechanical mechanisms and involves, for example, extrusomes called trichocysts and (b) the second is mediated by toxic secondary metabolites and involves different kinds of chemical extrusomes.

Keywords: protists, ciliates, extrusomes, secondary metabolites, chemical offense

1. Introduction

A common definition for predatory behavior describes it as the process through which one animal, the predator, captures and kills another animal, the prey, before eating it in part or entirely [1]; however, according to the opinion of a number of microbiologists and protistologists, this definition should be also extended to different organisms included in other life Kingdoms, with particular regard to microorganisms. Indeed, especially in the last 30 years, a lot of studies have been devoted to describing the predator-prey interactions among unicellular eukaryotic organisms like protists. Whittaker [2] originally defined protists as those "organisms which are unicellular or unicellular-colonial and which form no tissues," and for this reason they must carry out at the cellular level all the basic functions which can be observed in multicellular eukaryotes. Among these functions, self-nonself recognition mechanisms are represented by a large repertoire in protists and can trigger either autocrine or paracrine processes in some ciliates (see [3] for a review), together with the capability to detect prey (food) or predators in others. In this regard, it is known that protists have developed a variety of strategies of feeding behaviors especially in response to different environmental factors, together with a diverse kind of food available in micro-habitats. **Figure 1** shows a general scheme of predator-prey interactions, where the predator recognizes the presence of the prey (step 1) and can attack it (step 2). On the other hand the prey recognizes the presence of the predator (step 1′) and it can organize its defense mechanisms (step 2′) [4]. This scheme should be considered functional for both animals and protists, and indeed several studies have shown that the food recognition and the

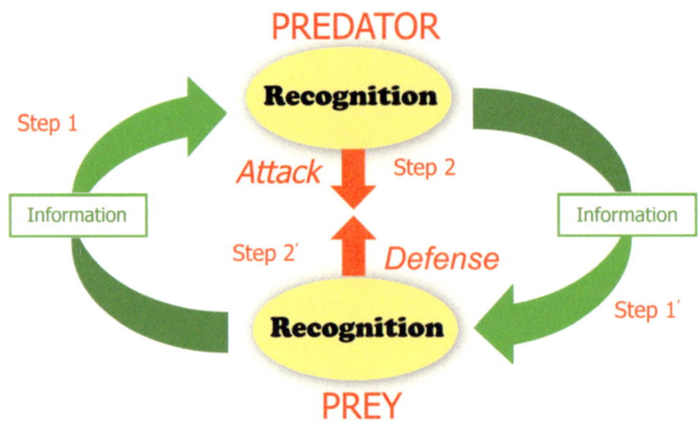

Figure 1.
General scheme of predator-prey interactions. Redrafted from [4].

offense-defense mechanisms adopted by some groups of protists can be compared, in terms of complexity and variability, with those observed in animals.

In this context, a common feeding mechanism found in heterotrophic protists is phagocytosis, a process which requires specific organelles for food assimilation and which occurs in three steps: food capture, phagosome formation, and food digestion [5]. Different techniques of phagocytosis have been described in various protists, where they have especially been investigated in ciliates [5–7]. Verni and Gualtieri [5] describe three main phagocytotic processes in ciliates: filter feeding, suctorial feeding, and raptorial feeding. The authors compare them to the strategies used in fishing, like netting, trapping, and harpooning. In filter-feeding ciliates, the food, represented by small organisms or edible debris of various types, was pushed into the ciliate buccal cavity by the rhythmical beats of the cilia located in its adoral apparatus. Suctorial-feeding ciliates are represented by sessile or sedentary species that for most of their lives remain attached to other organisms or various substrates, intercepting the food particles with their specialized tentacles. Finally, raptorial ciliates are able to directly catch other organisms using peculiar organelles to paralyze and/or kill their prey, generally called extrusomes.

2. Extrusomes, the specialized organelles for predator-prey interaction

The term "extrusome" was proposed, for the first time, by Grell in 1973 for extrusive (ejectable) bodies, which occur widely in protists [8]. They are membrane-bound organelles usually located in the cell cortex, attached to the cell membrane. They can display differences in structure and morphology, but they share the common characteristic of discharging their contents to the outside of the cell in response to mechanical or chemical stimuli. Remarkably, when the extrusomes are discharged, the cell remains intact and functional. Studies on extrusomes and related organelles have been reviewed by Hausmann [9], Dragesco [10], Kugrens et al. [11], Hausmann and Hülsmann [12], and Rosati and Modeo [13]. Typical examples of these organelles include toxicysts, trichocysts, mucocysts, cortical, or pigment granules in ciliates and flagellates, haptocysts in suctorians, and kinetocysts in heliozoan actinopods. Some extrusomes are known to be related in predator-prey interactions, for example, to catch and kill the prey (such as toxicysts, haptocysts, kinetocysts, and some cortical granules), or used as defensive organelles (such as the trichocysts and various cortical or pigment granules), but the role of other kinds of extrusomes such as the mucocysts in *Tetrahymena* or the trichites in Strombidiidae [13] still remains obscure.

3. Offensive extrusomes

Offensive extrusomes generally possessed by raptorial protists and located usually at or near the feeding apparatus are discharged after contact with a possible prey, which is immobilized, damaged, or firmly bound to the predator. Among these, organelles, certainly the most widely studied, belong to the category of toxicysts (toxic extrusomes) and they play an essential role in capturing and killing prey [7, 13]. Toxicysts are synthesized in Golgi or ER vesicles and are usually localized in the cell cortex attached to the cell membrane. Most of them are observed in species belonging to the class Litostomatea and subclass Haptoria, but they are also present in other predatory ciliates. They are usually positioned in a specific region of the cell, near the oral apparatus, and generally in the first portion which contacts the prey during the raptorial feeding [13]. Independently of the specific differences in the morphology of the cytostome, the toxicysts are always present in an appreciable number, for example, in the genera *Didinium*, *Dileptus*, *Prorodon*, *Litonotus*, *Colpes*, *Homalozoon*, and many others. In resting position, the toxicysts appear generally as rod-like elements (**Figure 2**), and could be discharged in milliseconds, if exposed to an appropriate stimulus such as contact with a prey (**Figure 3**) [7]. In this case, the tubules of the toxicysts are suddenly introduced into the cytoplasm of the prey's body, like hypodermic needles, to inject the toxic material. Hausmann [7] reports essentially two ways by which the toxicysts may be discharged: in the first case, there is a fusion of the toxicyst's membrane with the plasma membrane, followed by the tubule discharge via evagination; in the second, observed in certain ciliate species, a telescopic discharge of the tubules was observed. During or near the end of the toxicysts' discharge, the toxic secondary metabolites were secreted by the tubules. It is worth noting that this mechanism of discharging toxic substances shows the structural and functional similarities that can be found between the toxicysts in ciliated protists and nematocysts in Cnidaria, despite the substantial differences in size [7].

In contrast with recent and less recent studies about the nature of the toxic secondary metabolites used by ciliates in chemical defense, no exhaustive data are yet available about the composition of the toxins stored in the toxicysts of predatory ciliates. This is essentially due to the difficulty in separating the content of extrusomes from other molecules produced by the ciliate, in order to purify them at homogeneity for subsequent chemical and structural analyses.

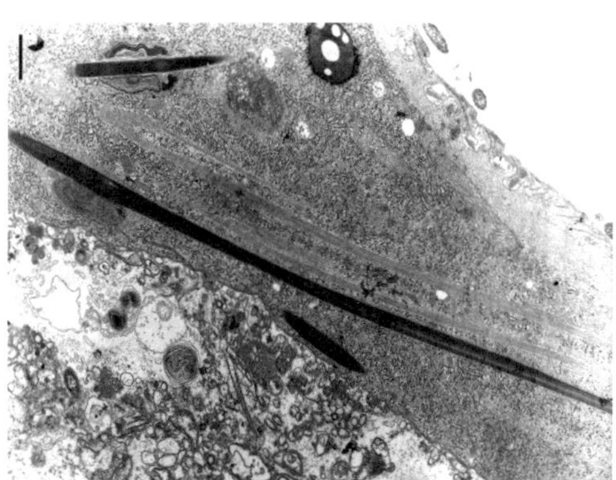

Figure 2.
Transmission electron microscope (TEM) picture of the toxicysts in a dividing cell of a ciliate Didinium nasutum. *Scale bar = 1 µm. Original picture by R. Allen from http://www.cellimagelibrary.org/images/10010.*

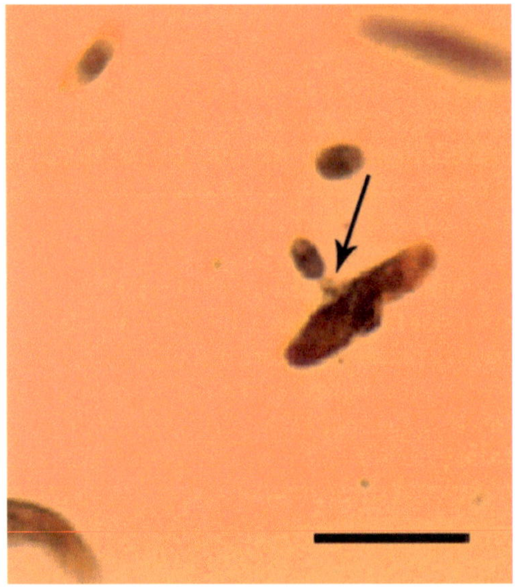

Figure 3.
Predatory behavior of Coleps hirtus *on* Pseudokeronopsis erythrina*. The predator attacks the prey with its toxicysts (arrow). Micrograph extracted from a film clips. Scale bar = 200 µm.*

To date the presence of acid phosphatase has been demonstrated in the toxicysts of *Didinium nasutum* [14] and four other raptorial ciliates such as *Enchelys mutans*, *Lacrymaria olor*, *Homalozoon vermiculare*, and *Pseudoprorodon niveus* [15]. It has been supposed that this enzyme, generally present in lysosomes of animal cells, may probably be used by these ciliates to start the digestion of the prey.

3.1 The predatory behavior of *Coleps hirtus*

The complete analysis of the content of the toxicysts, together with observations of the predatory behavior, was also performed on another species, *Coleps hirtus*, a freshwater protostomatid ciliate. *C. hirtus* (40–65 × 20–35 µm) has an oral apparatus placed at the anterior end of the cell and its barrel-shaped body is covered by calcified armor arranged in plates. This ciliate is able to feed off bacteria, algae, flagellates, and ciliates, but it is also histophagous, that is, it feeds on living plant and animal tissue such as rotifers, crustaceans, and fish [16, 17]. *Coleps* is also reported to show a scavenger feeding on tissues of dead metazoans, such as *Daphnia*, *Diaphanosoma*, and chironomid larvae [18], as well as toward dead ciliates and dead specimens of its own species. *Coleps* is usually equipped with toxicysts used by the ciliate to assist its carnivorous feeding, and its predatory behavior has recently been analyzed against another ciliate species used as prey, *Euplotes aediculatus.* Observations conducted on a mixture of predators and prey showed several contacts between the specimens of *Colpes* and *Euplotes*, but only after 5–10 min did interactions between the anterior section of a predator with a specimen of *Euplotes* become effective. This time was probably essential for prey detection and recognition, followed by prolonged contact between predator and prey, generally ending with the rapid backward swimming of the latter which separated the two organisms. When the attacks became numerous some individuals of *Coleps* remained attached to their prey (**Figure 4**), which decreased their swimming speed and gradually stopped swimming. After 20–30 min, the prey was fragmented and eaten by several specimens of *Coleps*, and a similar predatory behavior was also observed using different ciliate species as prey [19]. On the contrary the toxicysts-deficient

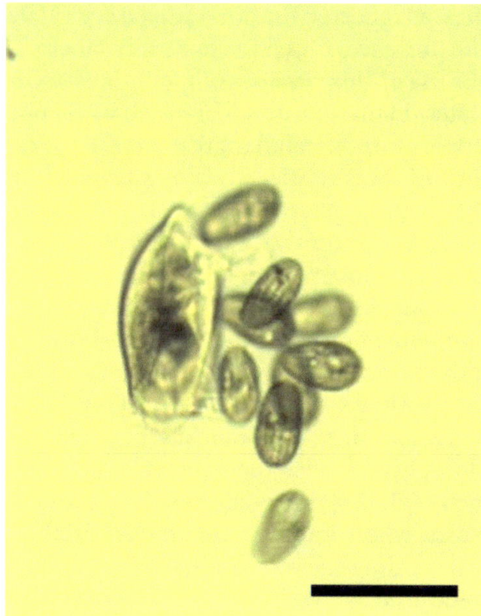

Figure 4.
Multiple attacks by different cells Coleps hirtus *on a cell of* Euplotes aediculatus. *Micrograph extracted from a film clips. Scale bar = 200 μm.*

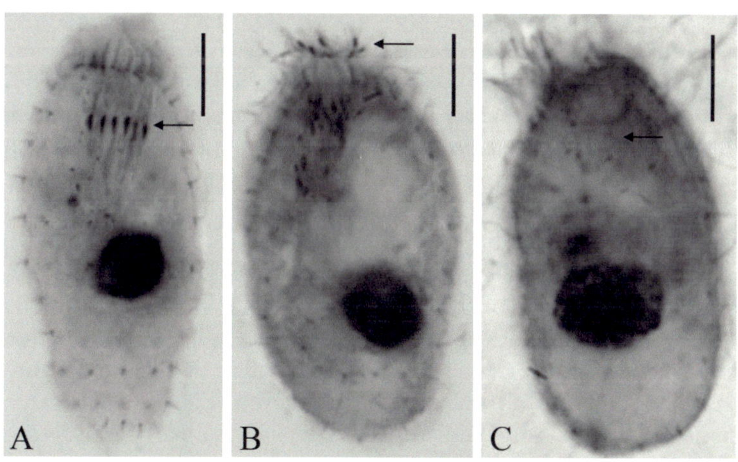

Figure 5.
(A) The toxicysts in Coleps hirtus *appear as rod-shaped organelles (arrow) in the oral basket of a cell. (B) The photomicrograph shows the toxicysts discharged (arrow) into the medium, immediately after a cold-shock treatment. (C) No toxicysts are detected in a toxicyst-deprived cell. Photomicrographs of fixed specimens by protargol stain. Scale bar = 10 μm. Pictures from [19].*

specimens of *Colpes* (**Figure 5**) obtained by means of the application of the cold-shock method capable of inducing an exclusive massive discharge of extrusomes in ciliates [20] appear unable to catch and kill their prey [19].

Unexpectedly, the analysis of the bioactive fraction of the toxicyst discharge of *Coleps hirtus* (performed by liquid chromatography-electro-spray-mass spectrometry and gas chromatography-mass spectrometry) showed the presence of a mixture of 19 saturated, monounsaturated and polyunsaturated free fatty acids (FFAs) with the addition of a minor amount of a diterpenoid (phytanic acid) but did not reveal the presence of enzymes, as reported for other predatory ciliates [19]. To date this is the only report on the presence of FFAs as toxic substances in the extrusomes of ciliated protists, but the use of this class of compounds as toxins by *Coleps* is shared with at least 15 freshwater, 13 marine, and 6 brackish water potentially harmful microalgae, as well with some multicellular organisms. For example, a chemical defense by a mixture of FFAs was studied and demonstrated for the harmful microalga *Fibrocapsa japonica* (Raphidophyceae) [21–23], and also in animals, a defensive strategy mediated by FFAs was recently described for the fish *Barbus barbus* which adopted it to protect its eggs from predators [24].

Very little is known about the role and source of phytanic acid in ciliates, this being the additional component detected in the toxicyst discharge of *Coleps*. Phytanic acid can be produced from the biodegradation of the side chain of chlorophyll [25], so one possible source arises from *Coleps'* carnivorous feeding on photosynthetic microorganisms [19]. Some insects, such as the sumac flea beetle, accumulate chlorophyll-derived metabolites as a chemical deterrent in excrements [26]. Komen et al. [27] demonstrated the toxic effect of phytanic acid on human skin fibroblasts, where it impaired mitochondrial respiration through protonophoric action. Regarding the role of phytanic acid in *Coleps*, it is possible to hypothesize that it can be used as a weapon, deterrent, or, at least, it could be stored in toxicysts given its potential toxic activity. In addition, it is known that ciliates themselves are also able to synthesize a huge number of terpenoids [28, 29]. This is the case of *Euplotes focardi* [30] and *Euplotes rariseta* [31] where the production of new diterpenoids was demonstrated. Terpene compounds and FFAs may also act together to exert cytotoxic effects [19]. FFAs may serve as a matrix to deliver toxic compounds to prey or predators and also to create a perfect environment where toxic metabolites can exert their functions.

It has been demonstrated that the substances discharged from the toxicysts by *Coleps* are highly toxic for a number of ciliate species such as *Euplotes aediculatus*, *Paramecium tetraurelia*, *Spirostomum teres*, and *S. ambiguum* or *Oxytricha* sp. [19], and their action mechanism appears to be related to a necrotic process. The term necrosis refers to a rapid (unprogrammed) cell death, with plasmatic membrane rupture, often caused by external factors such as toxins. On the contrary, the apoptosis is programmed cell death characterized by nuclear condensation, cytoplasmic shrinkage, and disintegration of the cell into small, membrane-bounded fragments. As shown in **Figures 6** and 7, the purified toxin from *Coleps* is able to induce rapid cell death in *E. aediculatus* and in *S. ambiguum* preceded by cell membrane fracture without any changes in the morphology of the macronucleus. An action mechanism of this type seems to be a "good choice" for *Coleps* as it induces paralysis and a very rapid death in the prey.

Interestingly, the cells of *Coleps* can also be damaged if exposed, *in vitro*, to their own toxin discharge [19]. Nevertheless, this cannot occur in nature, because on the one hand, the toxins are stored in the toxicysts of the ciliate, thus avoiding autotoxicity and on the other hand, the accidental exposure of *Coleps* to the toxicyst discharge dissolved in the medium is also unlikely, due to the choice of the predator to directly inject the toxins into the prey [19]. In this context, it is worth remembering the peculiar predatory behavior of *Coleps*, which usually leads to the observation that the same prey undergoes multiple attacks by several raptorial specimens, a behavior also adopted against young larvae of zebrafish [17]. It is likely that this behavior has evolved to ensure a fast immobilization of the prey, that after simultaneous multiple attacks, it can easily accumulate lethal concentrations of toxins injected by numerous predators. Therefore, essentially for the "wolf-like" group hunting behavior of *Coleps*, the species that appeared relatively resistant to its toxicyst discharge may also be easily caught and killed.

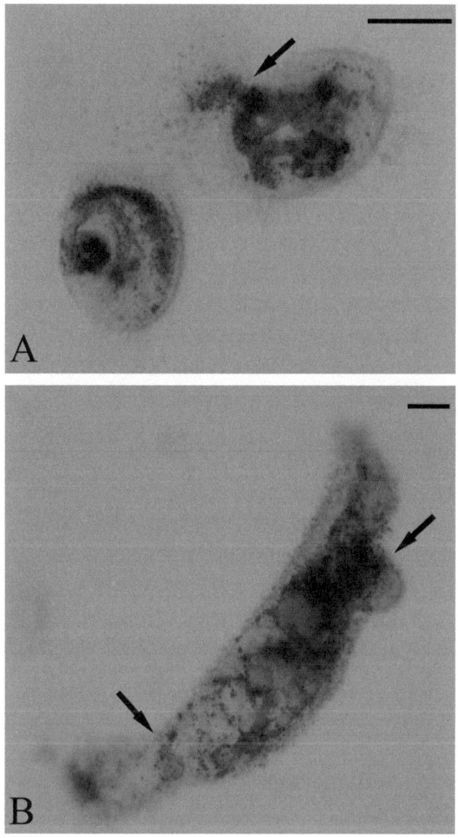

Figure 6.
(A) Necrotic effects of the toxicyst discharge of Coleps hirtus *on* Euplotes aediculatus *and (B)* Spirostomum ambiguum. *Arrows indicate the cell-membrane fractures. Scale bar = 100 µm. Pictures from [19].*

3.2 *Didinium nasutum*, a specialized hunter

Differently to *Coleps*, other ciliate species have specialized in hunting and catching a few preferential prey. This is, for example, the case of *Didinium nasutum* that is capable of capturing and killing several species of *Paramecium* and few other ciliates. Generally, *Paramecium* species are able to defend themselves by means of mechanical extrusomes like trichocysts (that will be discussed later on this chapter) but *Didinium* seems to overcome the defense of *Paramecium* by means of a highly specialized combination of extrusomes. Present on the proboscis of *Didinium* are several units of two different kinds of extrusomes: toxicysts, as in other Litostomatea, and pexicysts, another specialized offensive extrusome observed only in this species [32]. These authors describe the discharge of pexicysts as the first response after the prey recognition [14], which is typically followed by the discharge of toxicysts. At the same time, the prey (generally a *Paramecium*) discharges its trichocysts which separate the two organisms, but the proboscis of *Didinium* remains attached to the prey by a tiny connection probably composed of a bundle of discharged pexicysts and toxicysts (**Figure 8**). Subsequently, the *Paramecium* will be reached again and captured by the predator. In the light of this observation, the pexicysts seem to act most by a mechanical function (as harpoon-like organelles) rather than with a chemical offense. This assumption is supported by the fact that another species of predatory ciliate, *Monodinium balbiani*, which is morphologically similar and phylogenetically close to *Didinium*, but without the presence of the

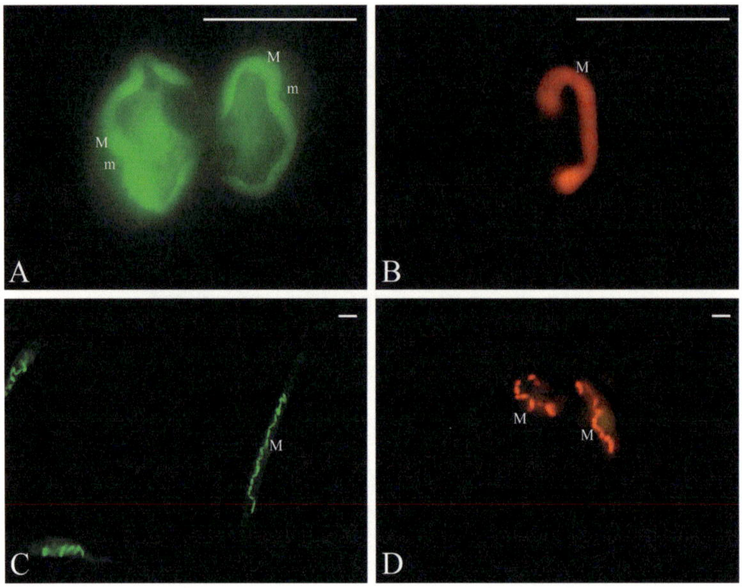

Figure 7.
(A, B) Effects of the toxicyst discharge of Coleps hirtus *on the macronuclear morphology in specimens of* Euplotes aediculatus *and (C, D)* Spirostomum ambiguum. *Cells were stained with acridine orange and ethidium bromide and observed by fluorescent microscopy. Viable cells show intact, bright green nuclei, nonviable cells show red/orange nuclei. M = macronucleus, m = micronucleus. Scale bar = 100 μm. Pictures from [19].*

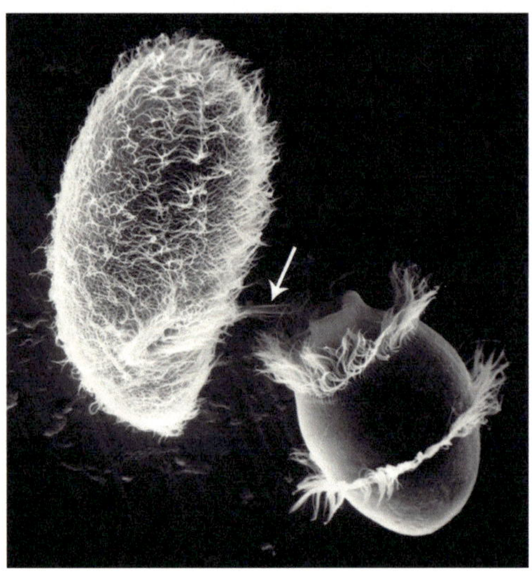

Figure 8.
Scanning electron microscope (SEM) picture on the predator-prey interaction between a cell of Didinium nasutum *and a cell of* Paramecium multimicronucleatum. *The bundle of toxicysts and pexicysts can be seen between the two organisms (arrow). Magnification ×50. Original picture by G. Antipa from http://www.cellimagelibrary.org/images/21991.*

pexicysts on its proboscis, unlike the *Didinium*, is sensitive to the defense mechanism possessed by *Paramecium,* which is often able to avoid capture [33].

3.3 The peculiar tentacles of suctorians

In this context it is also relevant to mention the subclass Suctoria, represented by ciliates which become sessile during development and consequently lose the ciliary structure. Suctorians are able to feed on other protists and frequently on other ciliates by means of specialized tentacles. The distal ends of these tentacles are often equipped by peculiar extrusomes called haptocysts that are involved in prey capture. When a tentacle touches a possible prey, the discharge of haptocysts is able to penetrate the prey's membrane, forming a connection between the predator and the prey and injecting the extrusome content into the latter, which also concurs to the fusion of the membranes belonging to the two organisms [13, 34]. However, the fusion of the two membranes is not always immediate, for example, in *Heliophrya erhardi,* Spoon et al. [35] observed that many specimens of *Paramecium* contacting the tentacles of the suctorian escaped discharging trichocysts at the point of contact, suggesting that *Paramecium* is able to defend itself from the puncture of the haptocysts.

4. Defensive extrusomes

In addition to predatory behavior, ciliated protists have also evolved different defense strategies, many based on the discharge of extrusomes. Two different mechanisms involved in their defense behavior are essentially observed: the first is mediated by the mechanical actions of trichocysts as in *Paramecium* or *Frontonia and* the second is mediated by the toxic secondary metabolites of different kinds of chemical extrusomes.

4.1 The mechanical defense

Spindle trichocysts (or simply, trichocysts) are spindle-shaped organelles which discharge their content in the form of a thread. They are found in some ciliates and flagellates and are sometimes furnished with a specially constructed tip [9]. The best known and studied trichocysts are those in the genus *Paramecium*. Trichocysts in *Paramecium* are 3–4 μm long, carrot-shaped membrane-bounded organelles armed with a sharply pointed tip, and are present in thousands all over the cell surface, except at the oral apparatus (**Figures 9** and **10**). When paramecia are subjected to various stimuli, the membranes of the trichocysts and the cell membrane blend together, and the content of the extrusomes is immediately discharged to the outside of the cell, forming a spear-like structure in milliseconds (**Figure 11**) (see [13] for a review). Trichocyst discharge has therefore been extensively studied as a model system of exocytosis [36] (see [37] for a review). Synthesis, processing, and sorting of component proteins in trichocysts are also studied as model systems of protein biosynthesis [36] for a review.

Maupas, one of the pioneers of protozoology, first proposed the defensive function of trichocysts in *Paramecium* in 1883, observing its morphological features and judging it as self-evident [38]; however, this point was questioned for years. The main controversy was due to the fact that *Paramecium* species are easily preyed upon by *Didinium* in spite of massive trichocyst discharge by paramecia. Pollack reported that *Didinium* preys on wild-type cells as easily as trichocyst-defective mutants in *P. tetraurelia* [39]. However, further studies have unequivocally indicated that trichocysts in *Paramecium* exert an effective defensive function against unicellular predators, including the raptorial protists *Dileptus margaritifer*, *Monodinium balbiani*, *Climacostomum virens*, *Echinosphaerium akamae,* and *E. nuceofilum* [33, 40–43]. In addition, a more recent paper also analyzed the defensive function of trichocysts in *P. tetraurelia* against some microinvertebrate predators, such as a rotifer (*Cephalodella* sp.), an ostracod (*Eucypris* sp.), and a turbellarian flatworm (*Stenostomum sphagnetorum*) [44]. The results of this study show the success in

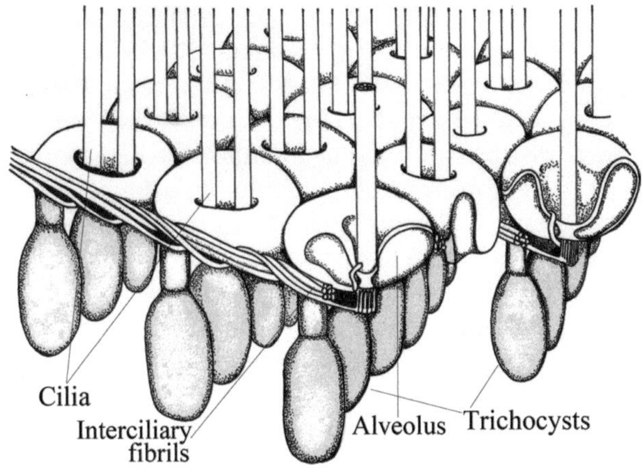

Figure 9.
Scheme of the ciliary structure and the trichocysts of Paramecium. *Picture from http://biodidac.bio.uottawa.ca, redrafted by R. D'Arcangelo.*

Figure 10.
Membrane details of resting trichocysts under the freeze fracture. The trichocyst tip (tt) and body (tb) are covered by the same membrane. The A-face of this membrane (A-tin) possesses randomly distributed particles whereas the B-face (B-tin) shows corresponding depressions. Scale bar = 1 μm. Picture from [9].

the defensive function of trichocysts against the rotifer and the ostracod while the mechanism seems ineffective against the flatworm. The authors speculate that the efficiency of the defense by means of trichocysts depends essentially on the kind of prey-capture behavior displayed by the predators. In particular, the success of the defense mediated by trichocysts appears positively related to the time that the predator requires to capture and manipulate the prey before ingestion. Consequently, and different from the turbellarian flatworm that directly swallows paramecia, predators such as the rotifer and the ostracod that, prior to ingesting paramecia, contact it with a ciliated corona or articulated appendices, give the prey sufficient time to activate the trichocysts discharge that allows it to escape [44]. Essentially this looks

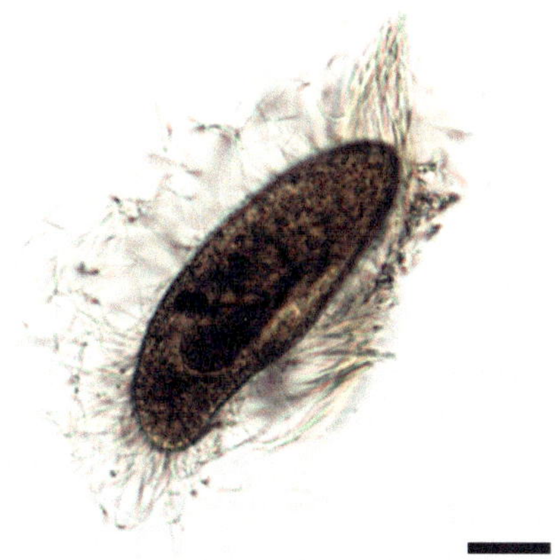

Figure 11.
The trichocysts discharged by a cell of Paramecium tetraurelia *exposed to picric acid solution. Scale bar = 100 µm.*

like the same phenomenon observed during the interaction between *Paramecium* and the predatory ciliate *Dileptus margaritifer*, that attempts to paralyze its prey with the toxicysts on its proboscis before ingestion, thereby inducing an explosive extrusion of trichocysts by *Paramecium*, which then swims away [44]. In this regard, another interesting observation was made when *Paramecium* was placed in a cell-free fluid containing the toxic material derived from the toxicysts from *Dileptus* [45] (Miyake A. personal communication); indeed after contact with this toxic solution, *Paramecium* cells violently reacted by immediately discharging most of their trichocysts before being killed. In this reaction, sometimes a single specimen (cell) of *Paramecium* was completely surrounded by its discharged trichocysts. When this occurred, the *Paramecium* survived long after other cells were killed, moving slowly in the narrow space in the capsule of discharged trichocysts. But when it happened that one of these encapsulated cells managed to squeeze out of the capsule, it was quickly killed. This observation suggests that discharged trichocysts of *Paramecium* function as a barrier against the *Dileptus* toxins and hence the locally discharged trichocysts in the *Paramecium-Dileptus* interaction function as an instant shield against *Dileptus*.

To summarize, the mechanical defense by trichocysts and related extrusomes appear to be multiple, including quick physical displacement, the thrust into a predator, and protection against the predator's toxins, increasing the chance for the prey to survive and escape. However, especially in ciliates and flagellates, other kinds of extrusomes used for defense were found, ones that, unlike trichocysts, are capable of discharging toxic materials in response to predatory behavior.

4.2 The chemical defense

Pigment granules (also called pigmentocysts) and **cortical granules** are extrusive organelles containing pigmented or colorless toxic material, respectively, and they were originally classified as a special type of mucocysts [9]. Pigment and cortical granules are mainly present in heterotrich and karyorelictean ciliates, such as *Blepharisma*, *Stentor*, *Loxodes,* and *Trachelonema,* but they may also exist in other groups of ciliates. They are usually present in great numbers throughout the cell cortex, sometimes providing bright colors to their bearers. Examples are *Stentor coeruleus*, whose coloration is due to the pigment called stentorin, and several red species of *Blepharisma*, whose coloration is due to blepharismins, formerly overall called zoopurpurin by Giese [46]. The coloration of these common heterotrichs has long attracted attention and most studies on pigment granules have been carried out using *S. coeruleus,* and a few red species of *Blepharisma*. *B. japonicum* (**Figure 12**) is the best studied species of the genus *Blepharisma* and it presents pigment granules usually in a size of 0.3–0.6 μm diameter, arranged in stripes between the rows of cilia that confer a red-pink coloration to the ciliate (**Figure 13**). These granules have been shown to contain a mixture of five compounds called blepharismins that are multifunctional quinone derivatives structurally related to hypericin, a photodynamic toxin of *Hypericum perforatum* (St. John's Wort), and stentorin, produced by the ciliate *S. coeruleus* [47, 48] (**Figure 14**). To date, two primary functions of blepharismins have been demonstrated: light perception and defense function against predators [47–52]. With regard to light perception, *B. japonicum* shows a temporal backward swimming or rotating movement (step-up photophobic response) if exposed to a sudden increase in light intensity. The step-up photophobic response helps the cells avoid strongly illuminated regions and lethal damage due to the photodynamic action of blepharismins [53]. In addition to light perception, blepharismins were found to act as chemical weapons via their light-independent cytotoxic effect against predatory protozoans and methicillin-resistant Gram-positive bacteria [49, 50, 54]. A possible explanation for this cytotoxicity can be found in the capability of blepharismins to form cation-selective channels in planar phospholipid bilayers [51], a phenomenon also expected to occur in the cell membranes of microorganisms exposed to toxic concentrations of ciliate pigments. The defensive function of blepharismins was initially proposed by Giese in 1949 who found that crude extracts of *Blepharisma* were toxic to various ciliates but not to *Blepharisma* itself [55]. Unfortunately, however, his preliminary tests did not support this assumption, that is, *Blepharisma* was easily eaten by predators such as the heliozoan *Actinospherium eichhorni* and small crustaceans [46, 55]. Some predators, *Didinium nasutum*, *Woodruffia metabolica,* and *Podophrya fixa*, did not

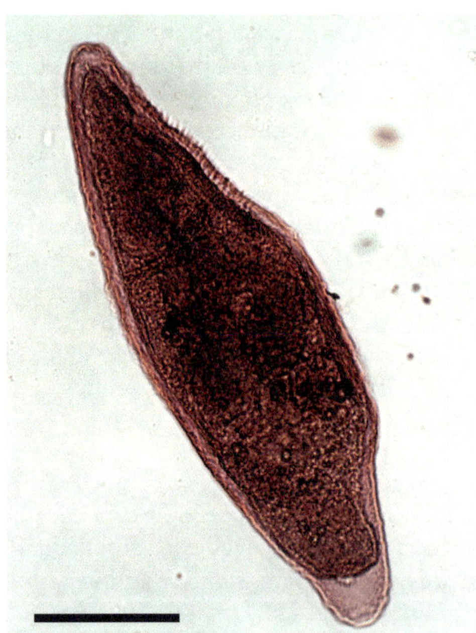

Figure 12.
External morphology of a living cell of Blepharisma japonicum. *Scale bar = 100 μm.*

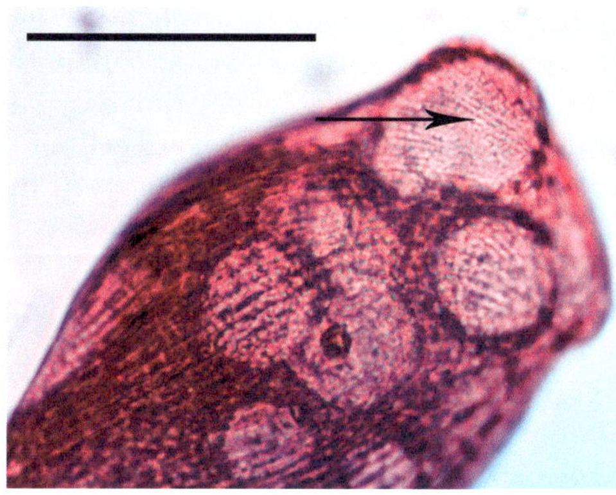

Figure 13.
Extrusive pigment granules in Blepharisma japonicum *(arrow) visible as red/pink dots under a vacuole. Scale bar = 100 μm.*

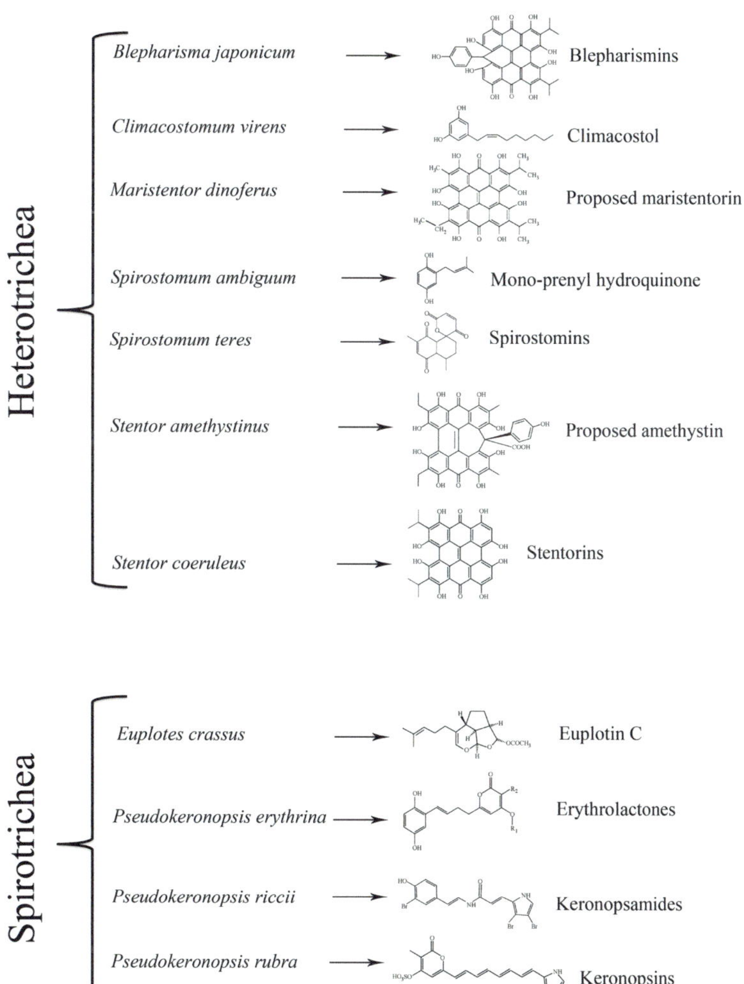

Figure 14.
Main secondary metabolites produced by ciliated protists. Erythrolactones: A1 ($R_1 = SO_3^-$; $R_2 = C_6H_{13}$ (n-hexyl)); B1 ($R_1 = SO_3^-$; $R_2 = C_7H_{15}$ (n-heptyl)); C1 ($R_1 = SO_3^-$; $R_2 = C_8H_{17}$ (n-octyl)); A2 ($R_1 = H$; $R_2 = C_6H_{13}$ (n-hexyl); B2 ($R_1 = H$; $R_2 = C_7H_{15}$ (n-heptyl)); C2 ($R_1 = H$; $R_2 = C_8H_{17}$ (n-octyl)).

eat *Blepharisma*, but they also ignored some other ciliates including uncolored ones. In the absence of further evidence, Giese was skeptical about the assumption [46]. This hypothesis was further unequivocally demonstrated by Miyake, Harumoto, and collaborators, comparing normally pigmented red cells of *B. japonicum*, albino mutant cells, and light-bleached cells (a phenocopy of the albino mutant) as prey for the raptorial ciliate *Dileptus margaritifer* and evaluating the toxicity of purified blepharismins on various ciliate species [49, 50]. As a response to the attack by *D. margaritifer* versus one cell of *B. japonicum*, the latter releases the toxic blepharismins, visible as spherical bodies of 0.2–0.6 μm in diameter under scanning electron microscopy (**Figure 15**). The discharge take place within a second and it is able to repel the predator, while the albino and light-bleached cells are much more sensitive to the attacks of *D. margaritifer* [49, 50]. Recently the defensive function of blepharismins was also investigated in two additional species of *Blepharisma, B. stoltei,* and *B. undulans* against two predatory protists (*C. hirtus* and *Stentor roeseli*) and one metazoan, the turbellarian *S. sphagnetorum* [56]. The results indicate that the chemical defense mechanism present in *B. stoltei* and *B. undulans* is mediated by the same five blepharismins present in *B. japonicum*, although produced in different proportions [56]. Authors speculate that the conservation of this panel

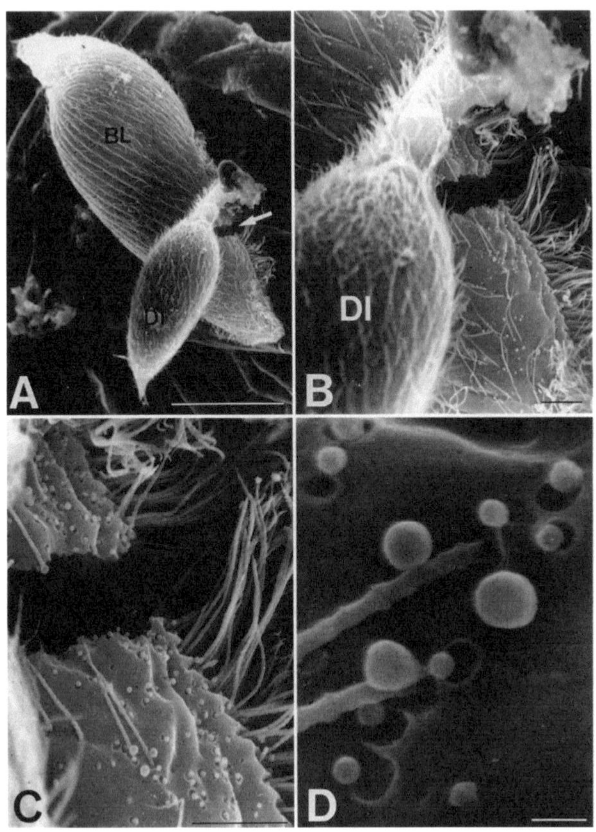

Figure 15.
SEM micrographs of the predator-prey interaction between a cell of Dileptus margaritifer *(DI) and a cell of* Blepharisma japonicum *(BL). (A)* Blepharisma *being attacked by* Dileptus. *Arrow indicates the site of the damage inflicted by the proboscis of the* Dileptus. *The rupture runs across the adoral zone of membranelles of the* Blepharisma. *Scale bar = 50 μm. (B) Enlargement of the region near the rupture in A. Scale bar = 5 μm. (C) The rupture magnification in B, showing the surface of* Blepharisma *peppered with spherules discharged from pigment granules. The surface is also pitted with small depressions presumably formed at the spots where the spherules have passed through the cell membrane. Scale bar = 5 μm. (D) Enlargement of a part of C. Scale bar = 0.5 μm. Pictures from [50].*

of toxic secondary metabolites suggests that distinct roles for these molecules are likely required at least for the fine control of photophobic reactions, as initially proposed by Matsuoka et al. [48]. Summarizing, the *Blepharisma* species studied are able to defend themselves against *C. hirtus,* although *S. sphagnetorum* and *S. roeseli* seem to overcome *Blepharisma*'s chemical defense, but it was observed that after the ingestion of intact cells of the toxic ciliates these predators are not able to reproduce, suggesting the presence of the post-ingestion toxicity phenomena [56]. Additional toxic pigments, structurally related to hypericin, were found in other heterotrich ciliate species, such as stentorin in *S. coeruleus* (see [57] for a review), amethystin in *S. amethystinus* [58], and maristentorin in the marine ciliate *Maristentor dinoferus* [59], but the defensive function was experimentally proved only for *S. coeruleus* [60].

Karyorelictean ciliates also possess pigment granules which are similar in size, structure, and distribution to those in the heterotrichs, but principally due to the difficulties to the growing species of karyorelictid in the laboratory, the chemical nature of their pigments is still unknown. The most studied species is freshwater *Loxodes striatus*, which presents yellow-brown pigment granules previously examined as photoreceptors [61]. More recently it has been proved that the pigment granules in *L. striatus* are extrusive organelles which contain a toxic photodynamic pigment used for chemical defense against predators [62]. *Loxodes* are able to discharge the toxic pigment as response to attacks of the ciliate *D. margaritifer* (**Figure 16**) or of the turbellarian *S. sphagnetorum* repelling predators. Intriguingly Finlay and Fenchel already proposed a defensive function for the pigment granules in *Loxodes* (*L. striatus* and *L. magnus*) based on different evidences; specifically,

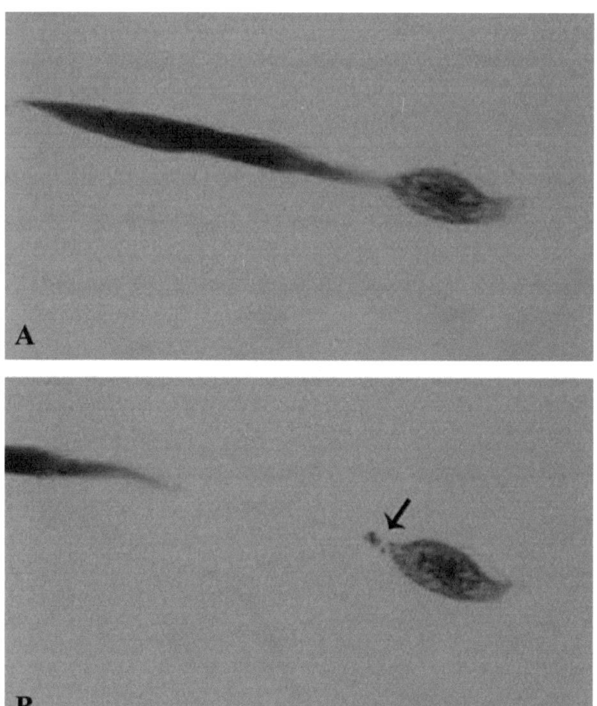

Figure 16.
Predator-prey interaction between Dileptus margaritifer *and* Loxodes striatus. *(A)* Dileptus *(the slender cell at the left) starts swimming backward after hitting a* Loxodes *with its proboscis. (B) The same cells as in A, about a second later, showing the retreated* Dileptus *and a mass of brownish material (arrow) near the* Loxodes. *Micrograph extracted from a film clips. Magnification ×70. Pictures from [62].*

they found that light induces in *Loxodes* a characteristic behavior to escape from toxic water and that the pigment granules are the photoreceptors for this reaction [61]. They assumed that this reaction may serve to localize *Loxodes* in regions of low oxygen tension where predators, such as planktonic metazoan, are rare and therefore the pigment may function as a predator-avoidance strategy. If this is the case, pigment granules of *Loxodes* participate in two very different kinds of defense, chemical defense and the behavior-based predator-avoidance, conferring to the ciliate an ability to defend itself against a wider range of predators [62].

Pigmented granules are found also in other groups of ciliates as the Spirotrichea, and mainly in the genus *Pseudokeronopsis,* which shows species equipped with reddish-brown pigment granules morphologically similar to those in heterotrichs [63]. Particularly in *P. carnea* [64] and in *P. erythrina* [65], these granules are reported as extrusive organelles. New secondary metabolites, keronopsins and keronopsamides, respectively, produced by *P. rubra* and *P. riccii,* were recently isolated together with their sulfate esters (**Figure 14**) [66, 67]. In the case of *P. rubra*, it was demonstrated that a crude extract of this organism containing keronopsins, A1 and A2, and their sulfate esters B1 and B2, is capable of paralyzing or even killing ciliates and flagellates [66]. For these reasons a defensive function for these secondary metabolites has been proposed; however, no data relative to their cellular localization and mechanism of action are available to date. On the other hand, in the case of *P. riccii*, the function of the alkaloid secondary metabolite keronopsamide A and its sulfate esters B and C has not been investigated, and the possible localization of the pigments in the cortical granules is only presumed [67]. The most extensively studied species is *P. erythrina*; previously described as an estuarine one, it was successively found also in the freshwater environment and hence reported as a euryhaline organism [68]. This ciliate shows an elongated body (**Figure 17**) equipped with spherical, dark-reddish, brown, or brick red colored pigment granules of about 1 μm in diameter that are mainly arranged around ciliary organelles [69]. As the content of pigment granules, three new secondary metabolites have recently been characterized and named erythrolactones A2, B2, and C2. These are characterized by a central 4-hydroxy-unsaturated δ lactone ring bearing an alkyl saturated chain at carbon-2 and a butyl-benzenoid group at carbon-5 [65, 68]. These molecules were detected in the crude extract of whole cells together with their respective sulfate esters, erythrolactones A1, B1, and C1 (**Figure 14**). After the application of the cold-shock method on massive cell cultures of *P. erythrina* to induce the exclusive discharge of pigment granules, it was demonstrated that only non-sulfonated molecules A2, B2, and C2 were contained in the extrusomes of the ciliate [65]. The mixture of these three molecules has been proven to repel some predators, such as the ciliate *C. hirtus*, and to be toxic for a panel of ciliates and microinvertebrates [65]. Erythrolactones A2, B2, and C2 are the only toxins present in the extrusome discharge of *P. erythrina*, whereas their respective sulfate esters

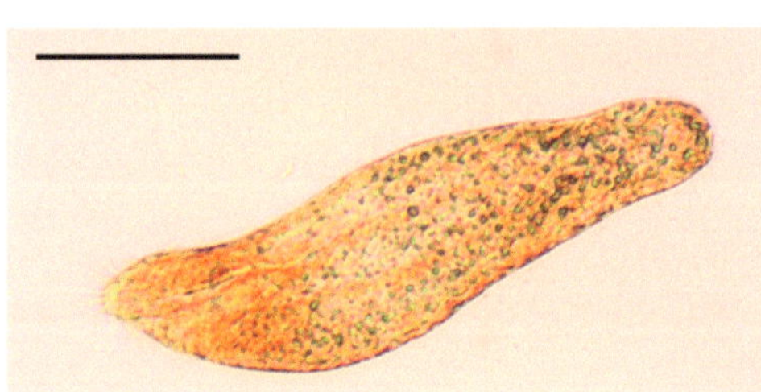

Figure 17.
External morphology of a living cell of Pseudokeronopsis erythrina. *Scale bar = 100 μm.*

A1, B1, and C1 remain confined inside the cell environment [68]. It is known that the process of sulfonation of endogenous molecules is a major metabolic reaction in eukaryotes that can increase water solubility and influence conformational changes but can also lead to the activation or inactivation of a biological effect (see [70] for a review). Buonanno and collaborators [64] speculate that the exclusive maintenance of the sulfate esters of the erythrolactones inside the *P. erythrina* cell may be associated with their temporary inactivation, in order to prevent the phenomenon of self-toxicity that could occur before their definitive storing, as non-sulfonated and active compounds, in the cortical pigment granules.

Other organelles strictly related to pigment granules are the colorless **cortical granules** in the heterotrich, sometimes reported as granulocysts to underline their extrusive nature. These organelles show a greatest morphological similarity to pigment granules, as in the case of the cortical granules of *Climacostomum virens* [71] and *Blepharisma hyalinum* [72]. The function and biological activity of the secondary metabolites contained in the cortical granules seem to be primarily related to chemical defense or offense, and the cortical granules in *C. virens* are to date the most studied. This freshwater heterotrich ciliate, if properly stimulated, is able to repel predators by discharging the colorless toxin climacostol (**Figure 14**) and some related analogues.

This toxin may be chemically classified within a large group of natural compounds known as resorcinolic lipids (also called alkylresorcinols or 5-alkylresorcinols), widely detected in prokaryotes and eukaryotes [73] and with reported antimicrobial, antiparasitic, antitumoral, and genotoxic activities (see [74] for a review).

A typical defensive behavior of *C. virens* occurs when a predator, such as the ciliate *D. margaritifer*, contacts a *C. virens* cell with its toxicysts bearing proboscis (**Figure 18A**). *D. margaritifer* swims backward while dense material is visible under dark field microscopy, emerging from the site where the proboscis touched the *C. virens* (**Figure 18B**) which swims away [75]. Sometimes, together with the discharged material from *C. virens*, it is also possible to detect some hazy material consisting of needle-like structures which appear to be discharged toxicysts of *D. margaritifer* (**Figure 19**), suggesting a possible further protection against the toxic extrusomes of predators [75]. Interestingly, the chemical defense adopted by *C. virens* against *D. margaritifer* is also effective against some other protists and metazoans [44, 76].

If the defensive function of cortical granules in *C. virens* is widely demonstrated, some evidences indicate that these extrusomes could be also successfully used for chemical offense. Differently from the *Paramecium* species which do not have trichocysts (exclusively for defense) localized in the oral apparatus, *C. virens* presents a wide number of cortical granules in the buccal cortex suggesting an additional offensive function for these extrusomes [71]. *C. virens* is able to catch and ingest prey of different sizes, from small flagellates such as *Chlorogonium elongatum* to large ciliates, such as *B. japonicum* or *Spirostomum ambiguum* [43, 77]. These prey are sucked up into the buccal cavity *of C. virens,* which is formed of a peristomial field and a buccal tube, and then ingested in a food vacuole, which arises at the end of the tube [43]. A cell of *P. tetraurelia* which is entirely taken into the buccal tube of *C. virens* is able to discharge the trichocysts and escape from the predator [43], different to what happens when an individual of the same species is totally caught in the pharynx of the microturbellarian *S. sphagnetorum* [44]. Perhaps, as in the case of contact with the toxicysts of the raptorial ciliate *D. margaritifer,* the trichocysts were discharged after contact with climacostol released from *C. virens* to kill the prey. A similar phenomenon also occurs with different preys which possess chemical extrusomes for defense such as the ciliate *S. ambiguum.* In this case, after a cell-cell contact, the *S. ambiguum* displays rapid cell contraction, and according to the authors, it is likely

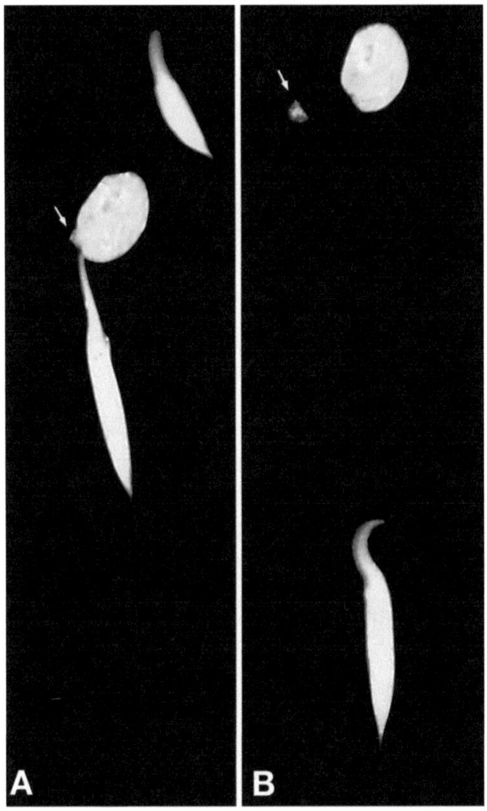

Figure 18.
Predator-prey interaction between Dileptus margaritifer *and* Climacostomum virens. *(A)* Dileptus *(the slender cell at the center) starts swimming backward after hitting with the proboscis* Climacostomum. *A small bulge (arrow) is developing on the surface of the* Climacostomum *at the site where the proboscis has just hit. (B) The same cells as in A, about a second later, show the retreated* Dileptus *and a small cloud (arrow) near the* Climacostomum. *Dark field micrographs of living cells. Magnification ×70. Pictures from [75].*

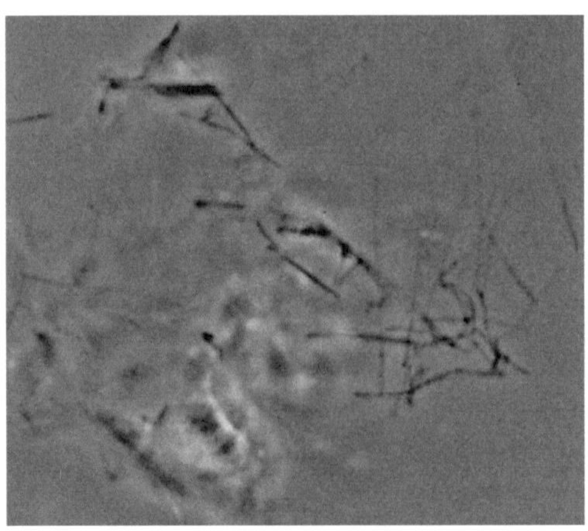

Figure 19.
Hazy cloud consisting of needle-like structures discharged from the toxicysts of Dileptus margaritifer. *Magnification ×720. Pictures from [75].*

that this contraction is induced by the discharge of extrusomes by *C. virens* [77]. If this is the case, it is likely that the cortical granules of *C. virens* could be equally used as multifunctional extrusomes, both for chemical defense and offense.

Besides the natural role of climacostol and thanks to the availability of a straightforward method for its chemical synthesis [78], other bioactivities of the toxin and its potential application to human health are, to date, investigated in various biological systems. The toxicity of climacostol proves very effective against pathogenic Gram-positive bacteria such as *Staphylococcus aureus* or *S. pneumoniae* and against a fungal pathogen, *Candida albicans* [79]. In addition, on the basis of the anticancer properties of other resorcinolic lipids, the toxic potential of climacostol is also studied against cancerous and non-cancerous mammalian cells, including human cell lines. The results show that climacostol effectively inhibits the growth of some tumor cell lines in a dose-dependent manner by inducing programmed cell death, with non-tumor cells proving significantly to be more resistant to the toxin [73, 80]. More recently the anti-tumor therapeutic activity of this toxin was also proved *in vivo*, using a melanoma allograft model in mice [81]. These results are quite interesting also in light of the fact that different molecules produced by other ciliate species show some particular pharmacological properties such as the sesquiterpenoid euplotin C or the cell type-specific signaling protein pheromone *Er-1* from *Euplotes* species (see [82] for a review).

Returning to the topic of this chapter, different secondary metabolites have been also isolated and characterized from other heterotrics, such as *Spirostomum ambiguum, and S. teres*. *S. ambiguum* (**Figure 20**) is a colorless freshwater species and one of the largest and elongated existing ciliates (800–2000 × 48–60 μm). The species is very common in the sludge-water contact zone of wells, ponds, sewage ponds, lakes, oxbows, ditches, and in the sediments of alpha- to beta-mesosaprobien rivers [77]. The defensive function of its cortical granules was recently demonstrated against different predators and the toxicity of its content was tested on a panel of freshwater ciliates [77, 83]. *S. ambiguum* has numerous cortical granules which, under a phase contrast microscope, appear as dots placed in the region between ciliary lines that could be observed in a large transparent contractile vacuole placed at

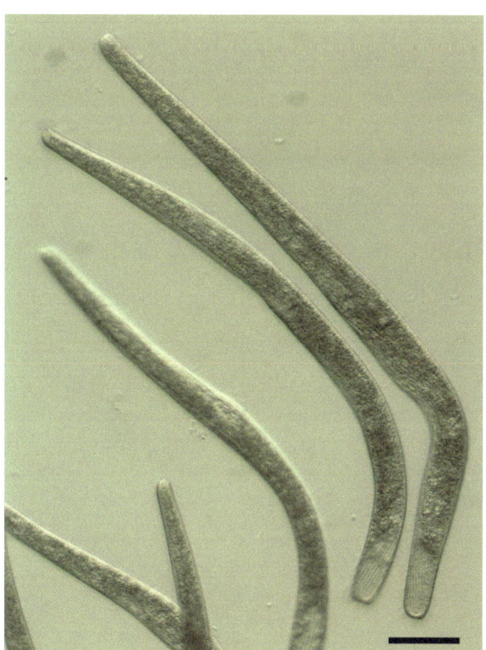

Figure 20.
External morphology of living cells of Spirostomum ambiguum. *Scale bar = 200 μm.*

the posterior end of the cell (**Figure 21A**) [77]. The cold-shock method was applied to *S. ambiguum* to obtain the cortical granule-deficient cells, which showed a markedly reduced number of extrusomes (**Figure 21B**). Both untreated and cortical granule-deficient cells were exposed to the attack of *C. virens*, and when the buccal apparatus of the predator makes contact with an untreated cell of *S. ambiguum,* it showed a rapid contraction while the predator swam backwards (**Figure 22A**). Similarly to untreated cells, cortical granule-deficient cells of *S. ambiguum* also showed rapid contraction after attack by *C. virens*, but they were successfully captured and sucked up by the predator into its buccal cavity (**Figure 22B**) [77]. The toxin involved in this interaction was purified by reversed phase high-performance liquid chromatography (RP-HPLC), and its structural characterization was carried out through nuclear magnetic resonance spectroscopy (NMR) and mass spectrometry (MS) measurements and revealed as 2-(3-methylbut-2-enyl)benzene-1,4-diol(mono-prenyl hydroquinone) (**Figure 14**). Prenylated-hydroquinone derivatives are metabolites of abundant occurrence and have been isolated from fungi, algae, plants, animals, and bacteria [77]. In this case the involvement of this molecule in predator-prey interaction is clear. Interestingly, another freshwater species of the genus *Spirostomum, S. teres,* possesses a different colorless toxin used for defense, characterized as spiro[(2,5-dimethyl-5,6,7,8-tetrahydronaphthalene-1,4-dione)-8,6′-(pyrane2',5′-dione)] and named spirostomin (**Figure 14**) [84]. It is no novelty that closely related organisms can produce different or even biogenetically distant specific secondary metabolites [77], and it is very common for ciliates [56]. To date, the only reported exception to this phenomenon is related to the genus *Blepharisma* in which the three species *B. japonicum, B. stoltei* and *B. undulans* share the same mixture of blepharismins even if produced in different proportions [56].

4.3 The inducible defense

Another peculiar defensive mechanism, reported as inducible defense, has been described for some *Euplotes* species as the response to the presence of some predators, such as microturbellarians, ciliates, or amoebas. These predators can release active substances, called kairomones, which induce some behavioral and morphological changes (such as the formation of spines in *Euplotes*) as a defensive mechanism in response to the presence of the predator [85–88] for a review.

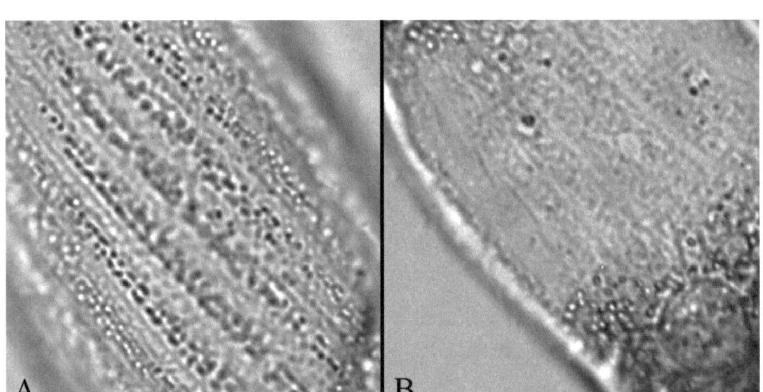

Figure 21.
Reduction in the number of extrusomes (cortical granules) in Spirostomum ambiguum *obtained by cold-shock treatment. (A) Extrusomes in an untreated cell. (B) Extrusome-deprived cell after cold shock. Magnification ×900. Pictures from [77].*

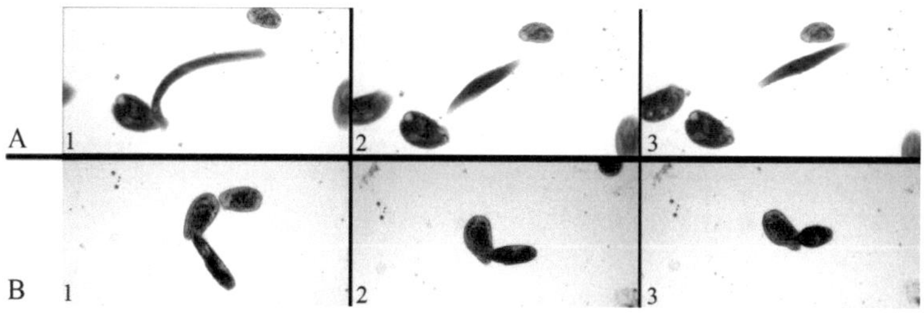

Figure 22.
Predator-prey interaction between Climacostomum virens *and* Spirostomum ambiguum. *(A) 1: Cell of* C. virens *contacts a cell of* S. ambiguum *with its buccal apparatus. 2:* S. ambiguum *shows rapid contraction while the predator swims backwards. 3: The same cells as in 2, a second later, showing a retreated* C. virens, *while* S. ambiguum *swims away. (B) Predator-prey interaction between* C. virens *and extrusome-deficient cells of* S. ambiguum *obtained by cold-shock treatment. 1:* C. virens *cell contacts a* S. ambiguum *cell which instantly shows contraction. 2:* C. virens *engulfs the contracted* S. ambiguum *cell and continues to eat the* S. ambiguum *cell (3). Micrographs extracted from a film clip. Magnification ×50. Pictures from [77].*

It could be interesting to study the efficiency of the inducible defenses, if compared to mechanical and chemical defense by means of extrusomes. In this regard, a first study was performed to compare the efficiency of the defense mediated by trichocysts in *P. aurelia* with that mediated by cortical granules in *C. virens* and *S. ambiguum* [44]. The authors reported that the mechanical defense in *Paramecium* against metazoan predators appears to be equally effective as the chemical one, but can be successfully activated only during the very early interactions with the predator, whereas it is ineffective after the ingestion of the ciliate. In contrast, the chemical defense adopted by a toxic ciliate against metazoan predators can also be activated after the ingestion of the prey by the predator, but its effectiveness appears to be strictly linked to the cytotoxic potency of the compound stored in the protozoan cortical granules. It would also be interesting to compare these two mechanisms against unicellular predators.

5. Conclusions

In a general perspective, it is clear that the researches on predatory behavior and on the related defensive mechanisms in protists not only represent progress in knowledge about the ecological role played in nature by predator-prey interactions in aquatic microhabitats but will also provide new research opportunities for evolutionary biology and may also represent a relevant source of new natural products.

Acknowledgements

We are grateful to Dr. Gill Philip (University of Macerata) for the linguistic revision of the chapter. Financial support was provided by University of Macerata, Italy.

Conflict of interest

The authors have declared no conflict of interest.

Author details

Federico Buonanno* and Claudio Ortenzi
Department of ECHT, Laboratory of Protistology and Biology Education,
University of Macerata, Macerata, Italy

*Address all correspondence to: federico.buonanno@unimc.it

References

[1] Minelli A. Predation. In: Jørgensen SE, editor. Encyclopedia of Ecology. 1st ed. Amsterdam: Elsevier B.V.; 2008. pp. 2923-2929

[2] Whittaker RH. New concepts of kingdoms of organisms. Science. 1969;**163**(3863):150-160. DOI: 10.1126/science.163.3863.150

[3] Luporini P, Alimenti C, Vallesi A. Ciliate pheromone structures and activity: A review. The Italian Journal of Zoology. 2015;**82**(1):3-14. DOI: 10.1080/11250003.2014.976282

[4] Harumoto T. Interazione cellulare interspecifica tra predatore e preda nei ciliati: Organelli e molecole che partecipano all'interazione [PhD thesis]. Italy: University of Camerino; 1993

[5] Verni F, Gualtieri P. Feeding behaviour in ciliated protists. Micron. 1997;**28**(6):487-504. DOI: 10.1016/S0968-4328(97)00028-0

[6] Radek R, Hausmann K. Phagotrophy of ciliates. In: Hausmann K, Bradbury PC, editors. Ciliates: Cells as Organisms. Stuttgart: Gustav Fischer Verlag; 1996. pp. 197-219

[7] Hausmann K. Food acquisition, food ingestion and food digestion by protists. Japanese Journal of Protozoology. 2002;**35**(2):85-95

[8] Grell KG. Protozoology. Berlin and New York: Springer-Verlag; 1973

[9] Hausmann K. Extrusive organelles in protists. International Review of Cytology. 1978;**52**:197-276. DOI: 10.1016/S0074-7696(08)60757-3

[10] Dragesco J. Capture et ingestion des proies chez les Infusories Ciliés. Bulletin Biologique de la France et de la Belgique. 1962;**96**:123-167

[11] Krugens P, Lee RE, Corliss JO. Ultrastructure, biogenesis and functions of extrusive organelles in selected non-ciliate protists. Protoplasma. 1994;**181**:164-190. DOI: 10.1007/BF01666394

[12] Hausmann K, Hülsmann N, editors. Protozoology. 2nd ed. New York: Thieme; 1996

[13] Rosati G, Modeo L. Extrusomes in ciliates: Diversification, distribution, and phylogenetic implications. Journal of Eukaryotic Microbiology. 2003;**50**:383-402. DOI: 10.1111/j.1550-7408.2003.tb00260.x

[14] Wessenberg H, Antipa G. Capture and ingestion of *Paramecium* by *Didinium nasutum*. Journal of Protozoology. 1970;**17**:240-270. DOI: 10.1111/j.1550-7408.1970.tb02366.x

[15] Fauré-Fremiet E. Pouvoir lytique et phosphatase acid chez le Ciliés. Comptes Rendus de l'Académie des Sciences. 1962;**254**:2691-2693

[16] Foissner W, Berger H, Scaumburg J, editors. Identification and Ecology of Limnetic Plankton Ciliates. München: Bayerisches Landesamt für WasserWirtschaft; 1999. pp. 272-287

[17] Mazanec A, Trevarrow B. *Coleps*, scourge of the baby Zebrafish. Zebrafish Science Monitor. 1998;**5**:1

[18] Auer B, Czioska E, Hartmut A. The pelagic community of a gravel pit lake: Significance of *Coleps hirtus viridis* (Prostomatida) and its role as a scavenger. Limnologica. 2004;**34**:187-198. DOI: 10.1016/S0075-9511(04)80044-6

[19] Buonanno F, Anesi A, Guella G, Kumar S, Bharti D, La Terza A, Quassinti L, Bramucci M, Ortenzi C. Chemical offense by means of toxicysts

in the freshwater ciliate, *Coleps hirtus*. Journal of Eukaryotic Microbiology. 2014;**61**(3):293-304. DOI: 10.1111/jeu.12106

[20] Buonanno F, Ortenzi C. Cold-shock based method to induce the discharge of extrusomes in ciliated protists and its efficiency. Journal of Basic Microbiology. 2016;**56**(5):586-590. DOI: 10.1002/jobm.201500438

[21] Fu M, Koulman A, van Rijssel M, Lützen A, De Boer MK, Tyl MR, Liebezeit G. Chemical characterisation of three haemolytic compounds from the microalgal species *Fibrocapsa japonica* (Raphidophyceae). Toxicon. 2004;**43**:355-363. DOI: 10.1016/j.toxicon.2003.09.012

[22] Pezzolesi L, Cucchiari E, Guerrini F, Pasteris A, Galletti P, Tagliavini E, Totti C, Pistocchi R. Toxicity evaluation of *Fibrocapsa japonica* from the Northern Adriatic Sea through a chemical and toxicological approach. Harmful Algae. 2010;**9**:504-514. DOI: 1016/j.hal.2010.03.006

[23] De Boer MK, Boerée C, Sjollema SB, de Vries T, Rijnsdorp AD, Buma AGJ. The toxic effect of the marine raphidophyte *Fibrocapsa japonica* on larvae of the common flatfish sole (*Solea solea*). Harmful Algae. 2012;**17**:92-101. DOI: 10.1016/j.hal.2012.03.005

[24] Mancini I, Defant A, Mesaric T, Potocnik F, Batista U, Guella G, Turk T, Sepcic K. Fatty acid composition of common barbel (*Barbus barbus*) roe and evalutaion of its haemolytic and cytotoxic activities. Toxicon. 2011;**57**:1017-1022. DOI: 10.1016/j.toxicon.2011.04.004

[25] Rontani J-F, Volkman JK. Lipid characterization of coastal hypersaline cyanobacterial mats from the Camargue (France). Organic Geochemistry. 2005;**36**(2):251-272. DOI: 10.1016/j.orggeochem.2004.07.017

[26] Vencl FV, Morton TC. The shield defense of the sumac flea beetle, *Blepharida rhois* (Chrysomelidae: Alticinae). Chemoecology. 1998;**8**:25-32. DOI: 10.1007/PL00001800

[27] Komen JC, Distelmaier F, Koopman WJH, Wanders RJA, Smeitink J, Willems PHMG. Phytanic acid impairs mitochondrial respiration through protonophoric action. Cellular and Molecular Life Sciences. 2007;**64**:3271-3281. DOI: 10.1007/s00018-007-7357-7

[28] Guella G, Skropeta D, Di Giuseppe G, Dini F. Structures, biological activities and phylogenetic relationships of terpenoids from marine ciliates of the genus *Euplotes*. Marine Drugs. 2010;**8**:2080-2116. DOI: 10.3390/md8072080

[29] Savoia D, Avanzini C, Allice T, Callone E, Guella G, Dini F. Antimicrobial activity of Euplotin c, the sesquiterpene taxonomic marker from the marine ciliate *Euplotes crassus*. Antimicrobial Agents and Chemotherapy. 2004;**48**(10):3828-3833. DOI: 10.1128/AAC.48.10.3828-3833.2004

[30] Guella G, Dini F, Pietra F. Epoxyfocardin and its putative biogenetic precursor, focardin, bioactive, new-skeleton diterpenoids of the marine ciliate *Euplotes focardii* from Antarctica. Helvetica Chimica Acta. 1996;**79**:439-448. DOI: 10.1002/hlca.19960790211

[31] Guella G, Callone E, Mancini I, Dini F, Di Giuseppe G. Diterpenoids from marine ciliates: Chemical polymorphism of *Euplotes rariseta*. European Journal of Organic Chemistry. 2012;**02012**:5208-5216. DOI: 10.1002/ejoc.201200559

[32] Wessenberg H, Antipa G. Studies on *Didinium nasutum*. Structure and ultrastructure. Protistologica. 1968;**4**:427447

[33] Miyake A, Harumoto T. Defensive function of trichocysts in *Paramecium* against the predatory ciliate *Monodinium balbiani*. European Journal of Protistology. 1996;**32**:128-133. DOI: 10.1016/S0932-4739(96)80048-4

[34] Benwitz G. Die Entladung der Haptocysten von *Ephelota gemmipara* (Suctoria, Ciliata). Zeitschrift für Naturforschung. Section C. 1984;**39**:812-817

[35] Spoon DM, Chapman GB, Cheng RS, Zane SF. Observations on the behavior and feeding mechanisms of the Suctorian *Heliophrya erhardi* (Rieder) Matthes preying on *Paramecium*. Transactions of the American Microscopical Society. 1976;**95**(3):443-462. DOI: 10.2307/3225137

[36] Adoutte A. Exocytosis: Biogenesis, transport and secretion of trichocysts. In: Gortz HD, editor. *Paramecium*. Berlin: Springer-Verlag; 1988. pp. 325-362

[37] Plattner H. Trichocysts *Paramecium*'s projectile-like secretory organelles: Reappraisal of their biogenesis, composition, intracellular transport, and possible functions. Journal of Eukaryotic Microbiology. 2017;**64**(1):106-133. DOI: 10.1111/jeu.12332

[38] Maupas E. Contribution a l'étude morphologique et anatomique des infusoires ciliés. Archives de Zoologie Expérimentale et Générale. 1883;**1**:427-664

[39] Pollack S. Mutations affecting the trichocysts in *Paramecium aurelia*: I. Morphology and description of the mutants. Journal of Protozoology. 1974;**21**:352-362. DOI: 10.1111/j.1550-7408.1974.tb03669.x

[40] Harumoto T, Miyake A. Defensive function of trichocysts in *Paramecium*. Journal of Experimental Zoology. 1991;**260**:84-92. DOI: 10.1002/jez.1402600111

[41] Knoll G, Haacke-Bell B, Plattner H. Local trichocyst discharge provides an efficient escape mechanism for *Paramecium* cells. European Journal of Protistology. 1991;**27**:381-385. DOI: 10.1016/S0932-4739(11)80256-7

[42] Harumoto T. The role of trichocyst discharge and backward swimming in escaping behavior of *Paramecium* from *Dileptus margaritifer*. Journal of Eukaryotic Microbiology. 1994;**41**:560-564. DOI: 10.1111/j.1550-7408.1994.tb01517.x

[43] Sugibayashi R, Harumoto T. Defensive function of trichocysts in *Paramecium tetraurelia* against heterotrich ciliate *Climacostomum virens*. European Journal of Protistology. 2000;**36**:415-422. DOI: 10.1016/S0932-4739(00)80047-4

[44] Buonanno F, Harumoto T, Ortenzi C. The defensive function of trichocysts in *Paramecium tetraurelia* against metazoan predators compared with the chemical defense of two species of toxin-containing ciliates. Zoological Science. 2013;**30**:255-261. DOI: 10.2108/zsj.30.255

[45] Miyake A. Cell-cell interaction by means of extru- somes in ciliates – Particularly on the predator–prey inteaction by extrusomal toxins. Japanese Journal of Protozoology. 2002;**35**:97-117

[46] Giese AC. *Blepharisma*. 1st ed. Stanford: Stanford University Press; 1973

[47] Lobban CS, Hallam SJ, Mukherjee P, Petrich JW. Photophysics and multifunctionality of hypericin-like pigments in heterotrich ciliates: A phylogenetic perspective. Photochemistry and Photobiology. 2007;**83**:1074-1094. DOI: 10.1111/j.1751-1097.2007.00191.x

[48] Matsuoka T, Kotsuki H, Muto Y. Multi-functions of photodynamic pigments in ciliated ptotozoans. In: Méndez-Vilas A, editor. Current Research, Technology and Education Topics in Applied Microbiology and Microbial Biotechnology. Badajoz: Formatex; 2010. pp. 419-426

[49] Miyake A, Harumoto T, Salvi B, Rivola V. Defensive function of pigment granules in *Blepharisma japonicum*. European Journal of Protistology. 1990;**25**:310-315. DOI: 10.1016/ S0932-4739(11)80122-7

[50] Harumoto T, Miyake A, Ishikawa N, Sugibayashi R, Zenfuku K, Iio H. Chemical defense by means of pigmented extrusomes in the ciliate *Blepharisma japonicum*. European Journal of Protistology. 1998;**34**:458-470. DOI: 10.1016/S0932-4739(98) 80014-X

[51] Muto Y, Matsuoka T, Kida A, Okano Y, Kirino Y. Blepharismins, produced by the protozoan, *Blepharisma japonicum*, form ion-permeable channels in planar lipid bilayer membranes. FEBS Letters. 2001;**508**:423-426. DOI: 10.1016/ S0014-5793(01)03110-6

[52] Uruma Y, Sakamoto K, Takumi K, Doe M, Usuki Y, Iio H. Assignment of ^{13}C NMR spectrum for blepharismin C based on biosynthetic studies. Tetrahedron. 2007;**63**:5548-5553. DOI: 10.1016/j.tet.2007.04.015

[53] Kato Y, Matsuoka T. Photodynamic action of the pigment in ciliated protozoan *Blepha-risma*. Journal of Protozoology Research. 1995;**5**:136-140

[54] Pant B, Kato Y, Kumagai T, Matsuoka T, Sugiyama M. Blepharismin produced by a protozoan *Blepharisma* functions as an antibiotic effective against methicillin-resistant *Staphylococcus aureus*. FEMS Microbiology Letters. 1997;**155**:67-71. DOI: 10.1111/ j.1574-6968.1997.tb12687.x

[55] Giese AC. A cytotoxin from *Blepharisma*. Biological Bullettin. 1949;**97**:145-149

[56] Buonanno F, Anesi A, Guella G, Ortenzi C. Blepharismins used for chemical defense in two ciliate species of the genus *Blepharisma*, *B. stoltei* and *B. undulans* (Ciliophora: Heterotrichida). European Zoological Journal. 2017;**84**(1):402-409. DOI: 10.1080/24750263.2017.1353145

[57] Song P-S, Kim I-H, Rhee JS, Huh JW, Florell S, Faure B, Lee KW, Kahsai T, Tamai N, Yamazaki T, Yamazaki I. Photoreception and photomovements in *Stentor coeruleus*. In: Lenci E, Ghetti E, Colombetti G, Hader D-P, Song P-S, editors. Biophysics of Photoreceptors and Photornovements in Microorganisms. New York: Plenum Press; 1991. pp. 267-279

[58] Höfle G, Reinecke S, Laude U, Kabbe K, Dietrich S. Amethystin, the coloring principle of *Stentor amethystinus*. Journal of Natural Products. 2014;**77**:1383-1389. DOI: 10.1021/np5001363

[59] Mukherjee P, Fulton DB, Halder M, Han X, Armstrong DW, Petrich JW, Lobban CS. Maristentorin, a novel pigment from the positively phototactic marine ciliate *Maristentor dinoferus*, is structurally related to hypericin and stentorin. Journal of Physical Chemistry. 2006;**110**:6359-6364. DOI: 10.1021/jp055871f

[60] Miyake A, Harumoto T, Iio H. Defensive function of pigment granules in *Stentor coeruleus*. European Journal of Protistology. 2001;**37**:77-88. DOI: 10.1078/0932-4739-00809

[61] Finlay BJ, Fenchel T. Photosensitivity in the ciliated protozoon *Loxodes*: Pigment granules, absorption and action spectra, blue light perception, and ecological significance. Journal of Protozoology.

1986;**33**(4):534-542. DOI: 10.1111/j.1550-7408.1986.tb05658.x

[62] Buonanno F, Saltalamacchia P, Miyake A. Defense function of pigmentocysts in the karyorelictid ciliate *Loxodes striatus*. European Journal of Protistology. 2005;**41**:151-158. DOI: 10.1016/j.ejop.2005.01.001

[63] Song W, Warren A, Roberts D, Wilbert N, Li L, Sun P, Hu X, Ma H. Comparison and redefinition of four marine coloured *Pseudokeronopsis* spp. (Ciliophora: Hypotrichida), with emphasis on their living morphology. Acta Protozoologica. 2006;**45**:271-287

[64] Wirnsberger E, Hausmann K. Fine structure of *Pseudokeronopsis carnea* (Ciliophora, Hypotrichida). The Journal of Eukaryotic Microbiology. 1988;**35**:182-189. DOI: 10.1111/j.1550-7408.1988.tb04321.x

[65] Buonanno F, Anesi A, Di Giuseppe G, Guella G, Ortenzi C. Chemical defense by erythrolactones in the euryhaline ciliated protist, *Pseudokeronopsis erythrina*. Zoological Science. 2017;**34**:42-51. DOI: 10.2108/zs160123

[66] Höfle G, Pohlan S, Uhlig G, Kabbe K, Schumacher D. Keronopsins A and B, chemical defence substances of the marine ciliate *Pseudokeronopsis rubra* (Protozoa): Identification by ex vivo HPLC. Angewandte Chemie International Edition. 1994;**33**:1495-1497. DOI: 10.1002/anie.199414951

[67] Guella G, Frassanito R, Mancini I, Sandron T, Modeo L, Verni F, Dini F, Petroni G. Keronopsamides, a new class of pigments from marine ciliates. European Journal of Organic Chemistry. 2010;**3**:427-434. DOI: 10.1002/ejoc.200900905

[68] Anesi A, Buonanno F, Di Giuseppe G, Ortenzi C, Guella G. Metabolites from the eury-haline ciliate *Pseudokeronopsis erythrina*. European Journal of Organic Chemistry. 2016;**7**:1330-1336. DOI: 10.1002/ejoc.201501424

[69] Chen X, Clamp JC, Song W. Phylogeny and systematic revision of the family Pseudokeronopsidae (Protista, Ciliophora, Hypotricha), with description of a new estuarine species of *Pseudokeronopsis*. Zoologica Scripta. 2011;**40**:659-671. DOI: 10.1111/j.1463-6409.2011.00492.x

[70] Strott CA. Sulfonation and molecular action. Endocrine Reviews. 2002;**5**:703-732. DOI: 10.1210/er.2001-0040

[71] Peck R, Pelvat B, Bolivar I, de Haller G. Light and electron microscopic observation on the heterotrich ciliate *Climacostomum virens*. Journal of Protozoology. 1975;**22**:368-385. DOI: 10.1111/j.1550-7408.1975.tb05187.x

[72] Larsen HF, Nilsson JR. Is *Blepharisma hyalinum* truly unpigmented. Journal of Protozoology. 1983;**30**:90-97. DOI: 10.1111/j.1550-7408.1983.tb01039.x

[73] Buonanno F, Quassinti L, Bramucci M, Amantini C, Lucciarini R, Santoni G, Ortenzi C. The protozoan toxin climacostol inhibits growth and induces apoptosis of human tumor cell lines. Chemico-Biological Interactions. 2008;**176**:151-164. DOI: 10.1016/j.cbi.2008.07.007

[74] Stasiuk M, Kozubek A. Biological activity of phenolic lipids. Cellular and Molecular Life Sciences. 2010;**67**(6):841-860. DOI: 10.1007/s00018-009-0193-1

[75] Miyake A, Buonanno F, Saltalamacchia P, Masaki ME, Iio H. Chemical defence by means of extrusive cortical granules in the heterotrich ciliate *Climacostomum virens*. European Journal of Protistology. 2003;**39**:25-36. DOI: 10.1078/0932-4739-00900

[76] Buonanno F, Ortenzi C. The protozoan toxin climacostol and its derivatives: Cytotoxicity studies on 10 species of free-living ciliates. Biologia. 2010;**65**:675-680. DOI: 10.2478/s11756-010-0071-1

[77] Buonanno F, Guella G, Strim C, Ortenzi C. Chemical defence by mono-prenyl hydroquinone in a freshwater ciliate, *Spirostomum ambiguum*. Hydrobiologia. 2012;**684**:97-107. DOI: 10.1007/s10750-011-0972-1

[78] Fiorini D, Giuli S, Marcantoni E, Quassinti L, Bramucci M, Amantini C, Santoni G, Buonanno F, Ortenzi C. A straightforward diastereoselective synthesis and evaluation of climacostol, a natural product with anticancer activities. Synthesis. 2010;**9**:1550-1556. DOI: 10.1055/s-0029-1218695

[79] Petrelli D, Buonanno F, Vitali LA, Ortenzi C. Antimicrobial activity of the protozoan toxin climacostol and its derivatives. Biologia. 2012;**67**:525-529. DOI: 10.2478/s11756-012-0030-0

[80] Quassinti L, Ortenzi F, Marcantoni E, Ricciutelli M, Lupidi G, Ortenzi C, Buonanno F, Bramucci M. DNA binding and oxidative DNA damage induced by climacostol-copper(II) complexes: Implications for anticancer properties. Chemico-Biological Interactions. 2013;**206**:109-116. DOI: 10.1016/j.cbi.2013.08.007

[81] Perrotta C, Buonanno F, Zecchini S, Giavazzi A, Proietti Serafini F, Catalani E, Guerra L, Belardinelli MC, Picchietti S, Fausto AM, Giorgi S, Marcantoni E, Clementi E, Ortenzi C, Cervia D. Climacostol reduces tumour progression in a mouse model of melanoma via the p53-dependent intrinsic apoptotic programme. Scientific Reports. 2016;**6**:27281. DOI: 10.1038/srep27281

[82] Catalani E, Proietti Serafini F, Zecchini S, Picchietti S, Fausto AM, Marcantoni E, Buonanno F, Ortenzi C, Perrotta C, Cervia D. Natural products from aquatic eukaryotic microorganisms for cancer therapy: Perspectives on anti-tumour proprieties of ciliate biocative molecules. Pharmacological Research. 2016;**113**:409-420. DOI: 10.1016/j.phrs.2016.09.018

[83] Buonanno F. The changes in the predatory behavior of the microturbellarian *Stenostomum sphagnetorum* on two species of toxin-secreting ciliates of the genus *Spirostomum*. Biologia. 2011;**66**(4):648-653. DOI: 10.2478/s11756-011-0061-y

[84] Sera Y, Masaki ME, Doe M, Buonanno F, Miyake A, Usuki Y, Iio H. Spirostomin, defense toxin of the ciliate *Spirostomum teres*: Isolation, structure elucidation, and synthesis. Chemistry Letters. 2015;**44**:633-635. DOI: 10.1246/cl.150044

[85] Kuhlmann H-W, Heckmann K. Interspecific morphogens regulating prey–predator relationships in protozoa. Science. 1985;**227**:1347-1349. DOI: 10.1126/science.227.4692.1347

[86] Kuhlmann H-W, Heckmann K. Predation risk of typical ovoid and winged morphs of *Euplotes* (Protozoa, Ciliophora). Hydrobiologia. 1994;**284**:219-227. DOI: 10.1007/BF00006691

[87] Kush J. Induction of morphological changes in ciliates. Oecologia. 1993;**94**:571-575. DOI: 10.1007/BF00566974

[88] Laas S, Spaak P. Chemically induced anti-predator defences in plankton: A review. Hydrobiologia. 2003;**491**(1-3):221-239. DOI: 10.1023/A:1024487804497

Chapter 3

Microbial Ecology in the Atmosphere: The Last Extreme Environment

Ángeles Aguilera, Graciela de Diego-Castilla, Susana Osuna, Rafael Bardera, Suthyvann Sor Mendi, Yolanda Blanco and Elena González-Toril

Abstract

The atmosphere is an extreme environment where organisms are subject to low temperatures and high radiation. Many of the microorganisms detected there appear in resistant forms or show mechanisms of adaptation designed to withstand these extreme conditions. Airborne microorganisms may play an important role in the global climate system, biogeochemical cycling, and health. Dust storms are the atmospheric phenomenon that move more topsoil through the Earth's atmosphere, and numerous microorganisms attached to dust particles are thus transported. The Iberian Peninsula is periodically affected by this phenomenon as African dust frequently reaches southern Europe and the Mediterranean basin. There are numerous methods for sampling airborne microbes, but factors such as low biomass and high variability of the atmosphere render them not yet sufficiently efficient. Very few studies have been conducted directly in the atmosphere via sampling using airborne platforms. The National Institute for Aerospace Technology has two CASA C-212-200 aircraft that have been suitably modified to operate as airborne research platforms. These aircraft are a unique tool for the study of atmospheric microbial diversity and the different environments where they can be found. A study of the airborne microbial diversity in a Saharan dust event from four aerobiology sampling flights is provided in advance.

Keywords: extremophiles, aerobiology, airborne, aerial platforms, aircraft, dust storm, Saharan dust

1. Introduction

Extremophile organisms capable of growing in extreme conditions draw considerable attention since they show that life is robust and adaptable and help us understand its limits. In addition, they show a high biotechnological potential [1, 2]. Most of the best-characterized extreme environments on Earth are geophysical constraints (temperature, pressure, ionic strength, radiation, etc.) in which opportunistic microorganisms have developed various adaptation strategies. Deep-sea environments, hot springs and geysers, extreme acid waters, hypersaline environments, deserts, and permafrost or ice are some or the most recurrent examples of extreme environments [3]. However, the atmosphere is rarely thought of as an extreme habitat. In the atmosphere, the dynamics of chemical and biological interactions are very complex, and the organisms that survive in this environment must tolerate high levels of UV radiation, desiccation (wind drying), temperature (extremely low and high temperatures), and atmospheric chemistry (humidity, oxygen radicals, etc.) [4]. These factors turn the atmosphere (especially its higher layers) into one of the most extreme environments described to date and the airborne microorganisms into extremophiles or, at least, multiresistant ones [5].

It is known that airborne cells can maintain viability during their atmospheric residence and can exist in the air as spores or as vegetative cells thanks to diverse molecular mechanisms of resistance and adaptation [2, 6]. The big question is whether some of them can be metabolically active and divide. Bacterial residence times can be several days, which facilitate transport over long distances. This fact, together with the extreme conditions of the atmosphere, has led researchers to think for years that they do not remain active during their dispersion. However, recent studies strongly suggest that atmospheric microbes are metabolically active and were aerosolized organic matter and water in clouds would provide the right environment for metabolic activity to take place. Thus, the role played by microorganisms in the air would not only be passive but could also influence the chemistry of the atmosphere. In any case, only a certain fraction of bacteria in the atmosphere would be metabolically active [2, 7].

Despite recognizing its ecological importance, the diversity of airborne microorganisms remains largely unknown as well as the factors influencing diversity levels. Recent studies on airborne microbial biodiversity have reported a diverse assemblage of bacteria and fungi [4, 8–12], including taxa also commonly found on leaf surfaces [13, 14] and in soil habitats [15]. The abundance and composition of airborne microbial communities are variable across time and space [11, 16–19]. However, the atmospheric conditions responsible for driving the observed changes in microbial abundances have not been thoroughly established. One reason for these limitations in the knowledge of aerobiology is that until recently, microbiological methods based on culture have been the standard, and it is known that such methods capture only a small portion of the total microbial diversity [20]. In addition, because pure cultures of microorganisms contain a unique type of microbes, culture-based approaches miss the opportunity to study the interactions between different microbes and their environment.

Another limitation for the study of aerial microbial ecology at higher altitudes or in open ocean areas is the difficulty of repeated and dedicated use of airborne platforms (i.e., aircraft or balloons) to sample the air. Most studies to date on the atmospheric microbiome are restricted to samples collected near the Earth's surface (e.g., top of mountains or buildings). Aircraft, unmanned aerial systems (UASs), balloons or even rockets, and satellites could represent the future in aerobiology knowledge [5, 21, 22]. These platforms could open the door to conducting microbial studies in the stratosphere and troposphere at high altitudes and in open-air masses, where long-range atmospheric transport is more efficient, something that is still poorly characterized today. The main challenge in conducting these kinds of studies stems from the fact that microbial collection systems are not sufficiently developed. There is a need for improvement and implementation of suitable sampling systems for platforms capable of sampling large volumes of air for subsequent analyses using multiple techniques, as this would provide a wide range of applications in the atmospheric, environmental, and health sciences.

In aerobiology, dust storms deserve special mention. Most of them originate in the world's deserts and semideserts and play an integral role in the Earth system [23, 24]. They are the result of turbulent winds, including convective haboobs [25]. This dust reaches concentrations in excess of 6000 $\mu g\ m^{-3}$ in severe events [26]. Dust and dust-associated bacteria, fungal spores, and pollen can be transported thousands of kilometers in the presence of dust [9].

In this chapter, we approach the atmosphere as an extreme environment and make use of some advanced data from an example of an in situ study of the atmosphere: the analysis of bacterial diversity of the low troposphere of the Iberian Peninsula during an intrusion of Saharan dust using a C-212 aircraft adequately improved for aerobiological sampling.

2. Atmosphere, an extremophile environment

It is well known that there is a biota in the atmospheric air. The first study dates back to the nineteenth century, which speak about the presence and dispersion of microorganisms and spores in the atmosphere [27, 28]. Although the atmosphere represents a large part of the biosphere, the density of airborne microorganisms is very low. Estimates suggest that from the ground surface up to about 18 km above sea level (troposphere), there is less than a billionth of the number of cells

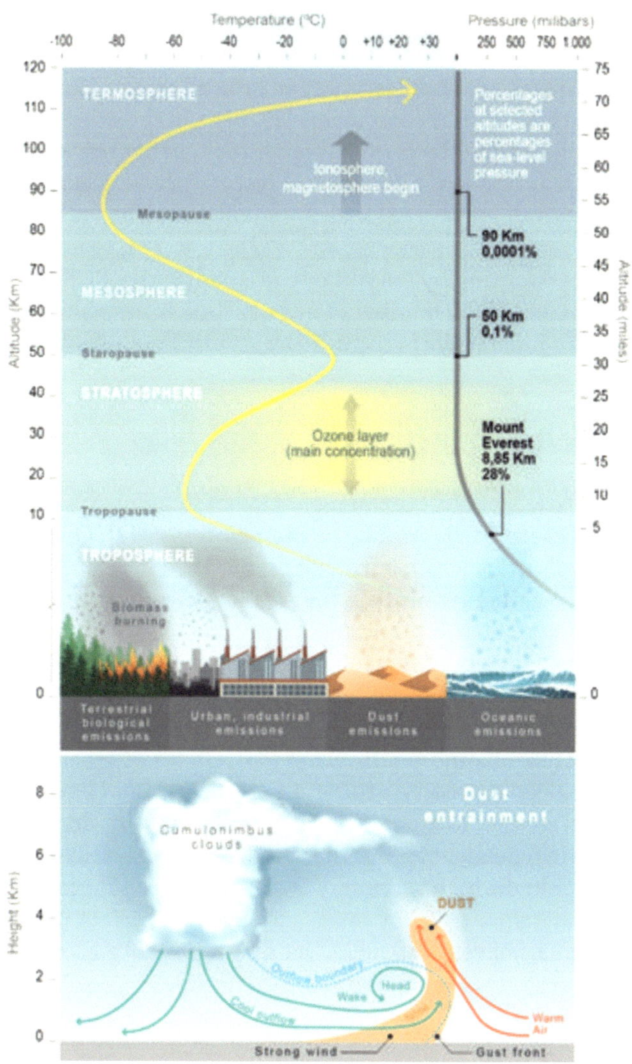

Figure 1.
Diagram displaying atmosphere layers, temperature and airborne emission sources. Yellow line marks atmospheric temperature. Bottom of the figures shows the common sources of aerosolized bacteria, with special attention to dust storms.

found in the oceans, soils, and subsurface. Between approximately 18 and 50 km above sea level (stratosphere), temperature, oxygen, and humidity decrease and with them the number of cells. Above the ozone layer (between 18 and 35 km into stratosphere), ultraviolet (UV) and cosmic radiation become lethal factors. Once in the mesosphere (above 50 km), life is difficult to imagine; however microorganisms of terrestrial origin could arrive to the stratosphere from lower layers via different phenomena (human activity, thunderstorms, dust storms, or volcanic activity), and bacteria have been found isolated up to 41 km or in dust samples from the International Space Station (**Figure 1**) [6, 29]. Therefore, airborne microbes are always present in the atmosphere [11, 30, 31], and their permanence is dynamic, resulting in an environment with enormous variability. Estimates calculate that over 1021 cells are lifted into the atmosphere every year, leading to considerable transport and dispersal around the atmosphere, with a large portion of these cells returning to the surface due to different atmospheric events as part of a feedback cycle. Undoubtedly, airborne microbes play an important role in meteorological processes. They have been linked to the nucleation phenomena that lead to the formation of clouds, rain, and snow and to the alteration of precipitation events [32–34]. Their presence is essential to understand long-range dispersal of plant and potential pathogens [7, 35, 36] and maintain diversity in ground systems and could interfere with the productivity of natural ecosystems [17, 18]. On the other hand, airborne bacteria can have important effects on human health, being responsible for different phenomena such as seasonal allergies and respiratory diseases. Based on data from terrestrial environments, the global abundance of airborne bacteria has been estimated to range between 104 and 106 m^{-3} [37]. However, more recent studies incorporating direct counting by microscopy or quantitative PCR have provided more accurate estimates of the number of airborne microbes, which apparently point to a higher number of cells present in the atmosphere [38–41].

3. Microbial sampling

There is a great variety of airborne microorganism sampling systems, allowing us to select the most suitable one depending on our objectives [42]. On the other hand, no standardized protocols exist, which is a major pitfall when developing our objectives. This fact has led some authors to propose the creation of consortiums of interested parties for establishing standardized protocol reproducibility [20], as well as the need to establish global networks of aerobiological studies [11]. Two approaches are proposed: particles or cells can be collected passively or directly from the atmosphere. Passive media usually involves decanting [43] and collecting particles over snow [44] or through the collection of atmospheric water [45]. On the other hand, active methodologies entail three major approaches: filtration, impaction, and liquid impingement. All three approaches are very efficient when developing culture-dependent techniques. In contrast, culture-independent approaches produce some serious problems that make the work difficult: the high variability of the system and the low biomass mean that sampling campaigns are, in many cases, extremely inefficient [20]. Lastly, the use of airborne platforms is not very extended, but they represent a good opportunity to conduct a more direct study of the atmosphere [5, 19, 31].

3.1 Filtration

Filtration is a simple and cheap method that is often efficient. It involves pumping air through a filter where the mineral and biological particles are trapped. Filters of different materials and porosity are available made of cellulose, nylon, polycarbonate or fiberglass, or quartz. Sizes used range from 0.2 to 8 μm, depending on the size of the particles to be captured and the capacity of the pump. In many cases, a PM10 filter can give better results when collecting smaller bacteria, as it allows greater airflow. Airflow filtration rates generally range between 300 and 1000 L/minute [4, 46]. Microorganisms trapped in the filter can be cultured, or the filters can be directly used for DNA extraction. In addition, filters are a very suitable support for microscopy, and countless holders for filters are available (an example is shown in **Figure 2A**).

3.2 Impingement

In impingement, particles are collected in a liquid matrix [20]. Normally a buffer is used such as phosphate buffer saline (PBS) that helps maintain the viability of the cells. One of the more widely used liquid impingers is BioSampler SKC (**Figure 2B**). In this case, the tangential movement of the particles inside the flow impinger retains the particles in the collecting liquid. The suspension obtained could be used for culturing or for molecular ecology assays [20]. One of the advantages of impingement collection is that it facilitates quantitative techniques such as flow cytometry or in situ hybridization [47].

3.3 Impaction

In this system, the particles generally impact into a petri dish with an enrichment medium. It is, possibly, the most efficient and most used method to conduct studies based on culture. Airflow impacting onto the plates is controlled by slots that allow the homogeneous distribution of the air. The system can be single stage or several stages in cascade, causing the particles to be distributed by size in the different petri dishes [20]. Some variants replace petri dishes with agarose filters or Vaseline strips, in order to carry out independent culture methodologies, but efficiency is very low. The original and more popular impactor is the Andersen cascade impactor (**Figure 2C**) [48].

3.4. Airborne platforms

Several studies explain and compare sampling methodologies in aerobiology, but most of them focus on the surface of the Earth (e.g., on top of mountains or buildings) or indoors [42, 49–54]. However, small studies have been conducted at higher altitudes or in open sea areas. The use of airborne platforms (balloons, aircraft, rockets, etc.) for aerobiology sampling would allow conducting a direct study of the microbial ecology of the atmosphere. Another advantage of airborne platforms is the possibility of studying the vertical distribution of airborne microbial communities. In addition, some aircraft allow us to develop studies in the upper troposphere or in the stratosphere. Unfortunately, atmospheric microbial collection instruments have not been developed enough for airborne platforms.

Among the different airborne platforms, aircraft, due to their versatility and access, are particularly interesting. Some studies have been conducted, but not enough samples have been developed yet, and efficiency is still very low. As already mentioned, the efficiency of samplers in soil-level aerobiology faces a series of problems (low biomass, high variability of populations, lack of standardized protocols). In the case of airplanes, in addition to these intrinsic problems associated with atmospheric microbial ecology, other additional ones exist: (1) the high velocity of the aircraft in relation to the relative quiescent air mass. This makes it difficult to obtain an isokinetic sampler and, therefore, one that is sufficiently efficient that would allow us to obtain a correct quantification of the incoming air [55]; (2) the sampler

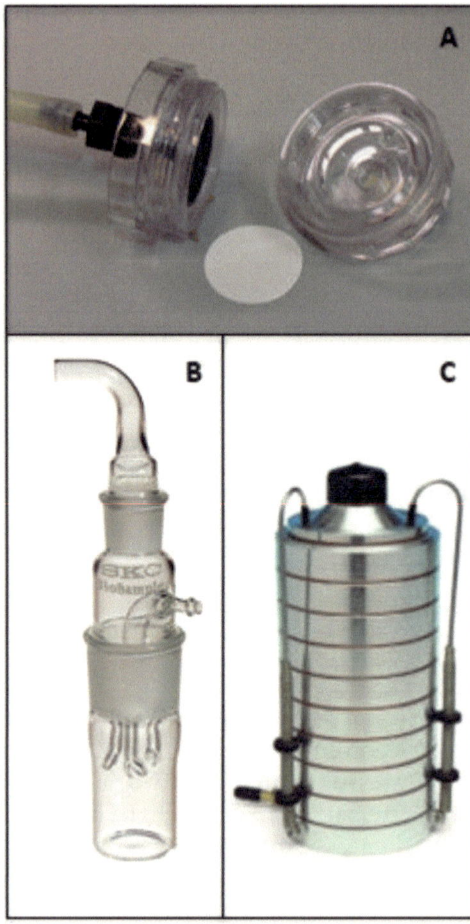

Figure 2.
Three different samplers of airborne microorganisms. (A) Filter holder and a filter (PALL Corporation). (B) Impinger sampling of bioaerosols (BioSampler, SKC, Inc.). (C) Six-stages Andersen Cascade Impactor (Thermo Fisher Scientific).

must be in a location on the airplane that avoids chemical contamination from the operation of the device. Previous studies have used wing-mounted air samplers or the roof of the aircraft to reduce the possibility of in-flight contamination [21, 22, 56–58]. Similarly, it should allow the aseptic collection of samples, avoiding microbiological contamination during the process. This operation, which can be very simple in the laboratory or at ground level, becomes tremendously complicated on an airplane, since air intakes that are part of the fuselage of the aircraft are often difficult to sterilize. It is therefore necessary to develop robust sterilization protocols. The spectacular work of DeLeon-Rodríguez of 2013 has been criticized in this aspect [40, 59]; (3) sampling time. A possible solution to the low biomass of the atmosphere is to increase sampling time, but in the case of flights, we are limited to the flight autonomy of the aircraft. Although scarce, some studies from airplanes have been conducted. The first studies that were conducted in airplanes were carried out by impaction on a petri plate with enrichment means, which allowed isolating microorganisms from the upper troposphere and even from the stratosphere [21, 57, 60]. However, advances in molecular ecology have caused the most recent studies to favor filtration [40, 58].

The European Facility for Airborne Research (EUFAR) program brings together infrastructure operators of both instrumented research aircraft and remote sensing instruments with the scientific user community. However, it lacked aircraft

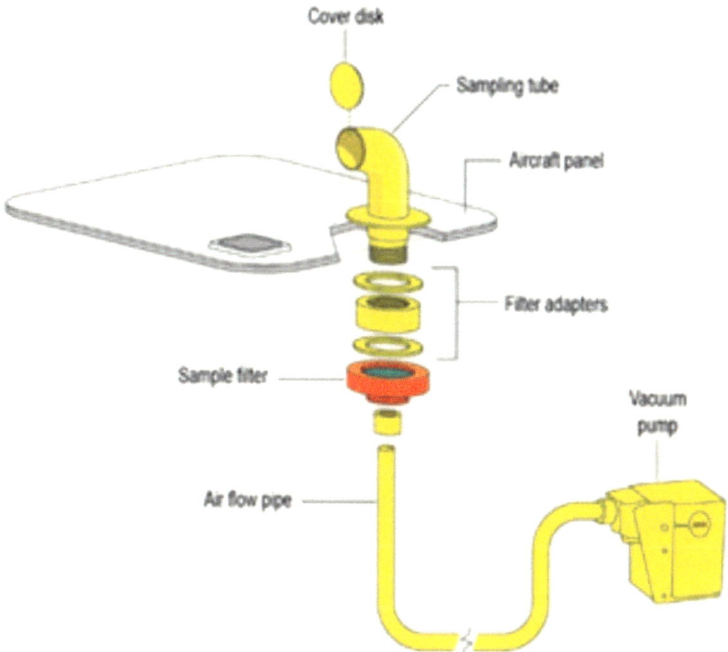

Figure 3.
Airborne microorganisms sampler installed in INTA's CASA C-212-200 aircraft.

Figure 4.
Multi-sampler system tested in INTA's CASA C-212-200 aircraft. (A) Impinger sampler, design and manufacture own. (B) Impactor sampler (Impaktor FH6, Markus Klotz GmbH). (C) Coriolis μ (Bertin Technologies SAS) a impinger biological air sampler. (D) Filter holder (PALL Corporation). (E) Six-stages Andersen Cascade Impactor (Thermo Fisher Scientific).

prepared for microbiological sampling. The National Institute for Aerospace Technology (INTA) belonging to the Spanish Ministry of Defence has two CASA C-212-200 aircraft that were suitably modified to be used as flying research platforms. Now, these two aircraft are a unique tool for the study of atmospheric microbial diversity and the different environments of the EUFAR program. Our research group has a CASA-212 aircraft with an air intake located on the roof of the aircraft. A metal tube fits the entrance and is fitted inside the aircraft to a filter holder, a flowmeter, and a pump (**Figure 3**). This simple system is easy to sterilize, and both the metal tube and the filter holder can be replaced in flight by other sterile ones if we want to take different samples. Using PM10 fiberglass filters, we can obtain isokinetic conditions and pass 1800 L of air per hour through the filter, as indicated by the flowmeter.

In a series of recent experiments, we tried to install a multi-sampler system in our aircraft, where we had five systems in parallel and connected to the same intake of the plane: one filter holder, two impingement systems, and two impactors (**Figure 4**). The results clearly showed that in the case of our aircraft, filtration was more efficient (data not shown).

4. Microbial characterization

Aerobiology studies have traditionally focused on the collection of bacterial cells and the analysis of samples by total counting and culture-based techniques. It is known that such methods capture only a small portion of the total microbial diversity [61]. The almost exclusive use, for years, of these methodologies is one of the reasons for these limitations in the knowledge of aerobiology. In addition, culture-dependent methods do not allow us to study the interactions between different species of microorganisms. Culture-independent methods have been used to assess microbial diversity, increasing the specificity of microbial identification and the sensitivity of environmental studies, especially in extreme environments. These methods have recently been applied to various areas of airborne microbiology [62–65] revealing a greater diversity of airborne microorganisms when compared to culture-dependent methods. Some good studies approach the challenges and opportunities of using molecular methodologies to address airborne microbiology [20, 66]. Although molecular ecology methods allow the rapid characterization of the diversity of complex ecosystems, the isolation of the different components is essential for the study of their phenotypic properties in order to evaluate their role in the system and their biotechnological potential. A combination of culture-dependent and culture-independent methods is ideal to address the complete study of the system.

Modern culture-independent approaches to community analysis, for example, metagenomics and individual cell genomics, have the potential to provide a much deeper understanding of the atmospheric microbiome. However, molecular ecology techniques face several particular challenges in the case of the atmospheric microbiome: (1) very low biomass [20]; (2) inefficient sampling methods [20]; (3) lack of standard protocols [9, 20]; (4) the composition of airborne microbes continuously changes due to meteorological, spatial, and temporal patterns [7, 62, 67–70]; and (5) avoidance of the presence of foreign DNA in the system [59]. Because these issues are not yet resolved, most of the non-culturing approaches focus on microbial diversity, where they are highly efficient.

The most recurrent techniques are those based on DNA extraction, gene amplification of 16S/18S rRNA, and next-generation sequencing (NGS) technologies. Often, this approach is more efficient due to the greater efficiency and sensitivity of this process, as opposed to gene cloning and Sanger sequencing; thus some authors are inclined toward metagenomics instead of amplification. This provides more information and avoids an intermediate step, but bioinformatic processing is tedious and often only provides data in relation to diversity, making the annotation of the rest of the information very complicated [20]. These approaches can be complemented with quantitative methods such as qPCR, flow cytometry, or fluorescence in situ hybridization (FISH) [41, 47, 66, 71]. FISH is surely the best and most specific cell quantification methodology that exists. However, in the case of aerobiology, it cannot always be used. A minimum number of cells must exist so that we can observe and count them under a fluorescence microscope. Due to the variability of microbial populations in the air, this is not always achieved. In our research group, we have obtained very good results in this regard, optimizing cell concentration. **Figure 5** shows epifluorescence micrographs of bacteria from an air sample. On this occasion, sampling was performed using a biological air sampler (Coriolis μ, Bertin Technologies SAS), where biological particles are collected and concentrated in a liquid (PBS). Sampling was conducted for 2 hours at ground level, pumping a total of 36,000 L of air. After this time, the sample was paraformaldehyde fixed and filtered through a 0.2 μm pore size, hydrophilic polycarbonate membrane, 13 mm diameter (GTTP, Millipore). A half sample was hybridized with the universal Bacteria domain probe, EUB338I-III [72], following a conventional protocol [73]. The second half was hybridized with the probe NON338 [74] as negative control. In this case, an average of 140 cells per liter of air was counted. Occasionally, FISH also allows to observe bacteria attached to mineral particles (**Figure 5C–D**).

DNA gives us much information about the diversity of the system, but if we wish to obtain information about the metabolic activity that is taking place in the ecosystem, metabolomic and metatranscriptomic approaches are needed [50, 66]. In the case of the atmosphere, this is crucial, since we are not fully certain if the cells present are active. Some studies indicate that a part of the microorganisms in the atmosphere are developing an activity [6], but until we conduct RNA-based and metabolite-based studies, we will not have the certainty that this is the case. The big problem is that it is very difficult to carry out these studies using the current microbial capture systems.

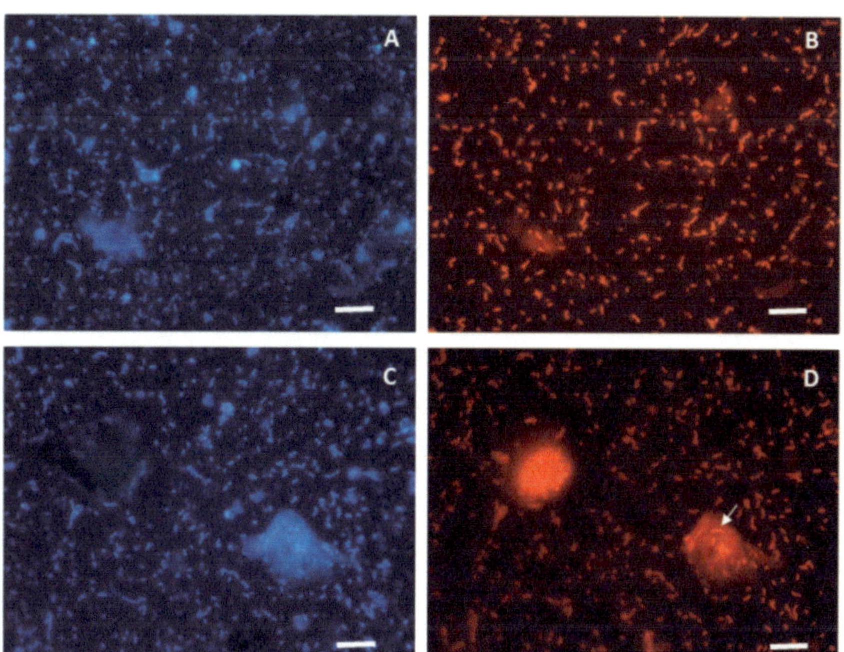

Figure 5.
Epifluorescence micrographs of bacteria from an air sample. (A and C) DAPI-stained cells; (B and D) same fields a A, and C, respectively, showing cells hybridized with probes EUB338I-III (Cy3 labeled), specific for Bacteria domain. All micrographs correspond to the same hybridization process, performed with a sample obtained after 4 hours sampling at ground. C and D show microorganisms attaches to a mineral particles (arrow sign). Bars, 5 μm.

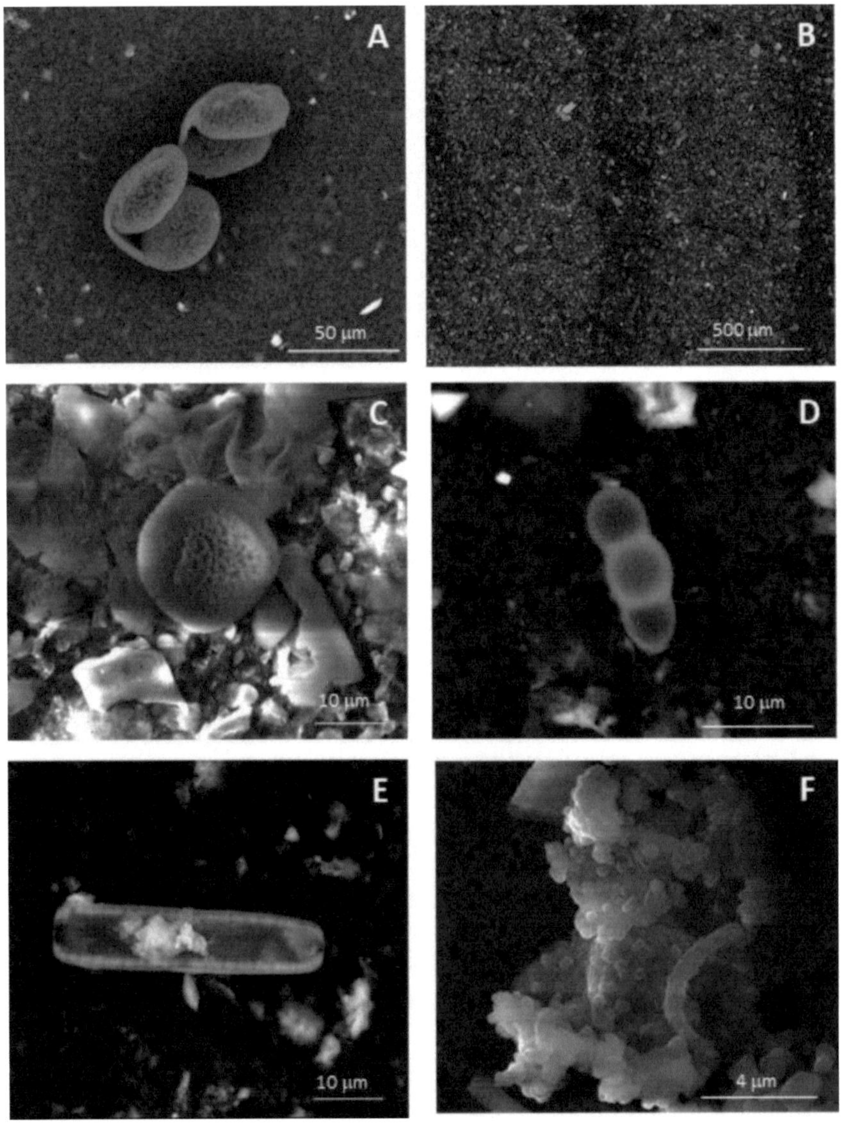

Figure 6.
SEM images of different airborne samples. (A) Pinus pollen. Ground sample after 2 hours sampling. (B) Air sample collected from C-212-200 aircraft during a Saharan dust intrusion (February 24, 2017). Filter appear completely cover of mineral particles. (B and C) Biological particles sampled using C-212-200 aircraft. (E) Diatomea sampled by C-212-200 aircraft in a fligth along the northern coast of Spain (9 March 2017). (F) Cell attached to mineral particles and organic matter.

Scanning electron microscopy (SEM) also provides much information of the aerobiology [7]. Specifically, it allows the characterization of eukaryotic cells (e.g., diatoms) and, above all, pollens and fungal spores, from which we can obtain great information with good images alone. **Figure 6A** shows pine tree pollen observed via SEM in a sample obtained after a 30 minutes flight of the C-212 aircraft.

5. Mechanisms of microbial survival of airborne bacteria

As mentioned above, factors, such as the shortage of nutrients and substrates, high UV radiation, drying, changes in temperature and pH, or the presence of reactive oxygen species, make the atmosphere an extreme environment. However, it is possible that the high variability of its conditions is the one characteristic that makes this environment more extreme [1, 20]. Among the cells present in the atmosphere, a considerable portion appears in the resistance forms capable of withstanding low-temperature and high-radiation conditions. This is what probably happens with fungi and gram-positive bacteria. *Bacillus* strains recurrently isolated from the atmosphere have characteristics and a capacity to sporulate very similar to strains isolated from the soil. Undoubtedly, another part of the cells will be in the form of latency and may even suffer modifications of the cell wall and slow down or stop their metabolic activity [75, 76]. These transformations can improve resistance to physical stresses, such as UV radiation [58]. On the other hand, some of the bacteria present in the atmosphere, such as *Geodermatophilus*, show pigmentation that undoubtedly protects it from excessive radiation. The microorganisms that are usually detected in the atmosphere originate mainly from the soil, which means they will share similar mechanisms of resistance. In some strains, metabolic adaptations have been observed to lack nutrients such as cytochrome bd biosynthesis to survive iron deprivation [77]. *Deinococcus* is also a recurrent genus in the atmosphere, which, like those in soil, has multiresistance mechanisms based on high DNA-repair efficiency. Bacteria that do not form spores and certain archaea, in contrast, often have genomes rich in G + C, which may increase tolerance to UV rays and overall survival [78].

Another strategy of resistance could be cell clustering and adhesion to particles. Several studies have confirmed the loss of viability and shielding or the reflective properties of the mineral particles as an important role for the protection of UV radiation [19, 31]. In that sense, it is very possible that many cells have mechanisms that promote aggregation. In our samples, we often find the cells adhered to each other or to minerals, which undoubtedly makes them more resistant (**Figure 6**).

6. Emission sources

Global and regional models have been used to explain bioaerosol emission, transport, and atmospheric impact [17, 18, 79–84]. Even so, it is not an easy phenomenon to explain, since it depends on a large number of factors. On the one hand, there are numerous sources of tropospheric aerosols, which include sea salt, volcanic dust, cosmic dust, industrial pollutants, and desert and semidesert areas [6, 85]. We must also consider the factors that make the transfer of particles possible, for example, meteorological phenomena, solar radiation, temperature, tides, erosion, etc. [85]. On the other hand, anthropogenic activities can also affect dust emissions indirectly, by changing the climate and the hydrological cycle. In these aerosols, microorganisms will be included in a greater or lesser number. The degree of richness in cells of tropospheric aerosols will depend largely on the source of emission. Thus, the large wooded masses or fields of crops provide the atmosphere with a good number of microorganisms due to the effect of air or the aerosols produced by rain. Similarly, anthropogenic activity contributes large amounts of bacteria to the environment, treatment plants, and composting areas being sources of airborne microorganisms [85].

Desert dust storms play a major role in particle emissions and with them that of microorganisms. In this way, most of the material reaching the atmosphere from the surface comes from desert and semidesert areas, which is known as desert dust. The Sahara-Sahel desert, the Middle East, central and eastern Asia, and Australia are the major sources of desert dust, although all the arid zones of the world are emission sources [9, 86]. Dust storms are atmospheric events typically associated with dry lands due to the preponderance of dried and unconsolidated substrates with little vegetation cover. The strong and turbulent winds that blow on these surfaces raise fine-grained material, a large part of which consists of particles the size of silt (4–62.5 μm) and clay (<4 μm), reducing visibility to less than 1 km. The atmospheric concentrations of PM10 dust exceed 15,000 μg/m^3 in severe events [87], although the concentrations naturally decrease with the distance from the areas of origin, extending hundreds of kilometers. The dust particles and cells associated with them are transported in this manner and will be deposited finally, by the effect of rain, snow, or other meteorological phenomena. Therefore, there is a continuous transfer of mineral and biological matter through the atmosphere that moves from the air to the terrestrial environment and changes its geographical area [7, 24].

7. Saharan dust

The Sahara-Sahel desert located in northwestern Africa is one of the major sources of windblown dust in the world [9]. This phenomenon has an impact on the Mediterranean coastline, but Saharan dust has been transported toward the north of Europe and has been found on numerous occasions in the Alps [88, 89] or blown toward the Atlantic and Caribbean [8, 90]. It has been estimated that 80–120 tons of dust are transported annually through the Mediterranean toward Europe [23, 91, 92]. In particular, dust transported by the winds can reach an elevation of up to 8 km in the atmosphere over the Mediterranean basin [93]. Because of its geographic position, the Iberian Peninsula is often affected by these dust events. Specifically, the Sahara-Bodele depression, located at the southern edge of the Sahara desert, has been described as the richest dust source reaching the Iberian Peninsula. Southern Spain is the main area affected, but dust can reach the Pyrenees and even France [43]. Different researchers have studied the mineralogical and chemical composition of Saharan dust, which has been observed to contain calcite, dolomite, quartz, different clay minerals, and feldspars as the main mineral components [94]. The intrusion of big amounts of these components is an important influence on nutrient dynamics and biogeochemical cycling in the atmosphere of the Iberian Peninsula.

Despite the large number of studies on dispersion, geochemistry, and mineralogy of African dust, few are focused on microbiology. All these studies conclude that there are microbes associated with dust because there are higher concentrations of aerosolized microorganisms during dust events [43, 90, 93–96]. However, the magnitude of the concentrations and the specific microbes associated with dust events remain the subject of debate. On the other hand, the viability of these microorganisms is another big question. The United States Geological Survey (USGS) develops the Global Dust Program to investigate the viability of microorganisms transported in dust masses. USGS authors using DNA sequencing of the ribosomal gene were able to isolate and identify more than 200 viable bacteria and fungi in St. John's samples in the USA [8, 36, 90]. Fungi and bacteria associated with atmospheric dust can be recovered and cultivated, but they must be gram-positive bacteria and many spore formers, which makes them resistant to the extreme conditions of the atmosphere.

Therefore, fungi and bacteria associated with dust may have been isolated from dust intrusions, but a percentage of the viable ones already remains an unanswered question. Another big question is the activity of these cells in the atmosphere. It is clear that they are resistant to extremophile conditions, but the question is whether they are developing their life cycle in this particular environment. This question could be answered by molecular ecology methodologies based on the isolation and sequencing of mRNA, but low atmospheric biomass and high variability are, once again, the great problem when developing this type of

RNA-based methodologies. On the other hand, clinical records point to many of the viable microorganisms identified in the Saharan dust as the cause of respiratory diseases (asthma and lung infections or allergic reactions), cardiovascular diseases, and skin infections [7, 90, 97, 98]. It is known that other microbes associated with dust in the air are pathogenic to humans, including those that cause anthrax and tuberculosis, or to livestock (such as foot and mouth disease) or plants [7, 90, 97, 98]. Characterization, quantification, and feasibility studies are vital to address these problems.

It is common to find fungal spores belonging to the genus *Aspergillus*, *Nigrospora*, *Arthrinium*, and *Curvularia* associated with Saharan dust. Bacterial taxa comprised a wide range of phyla, including *Firmicutes*, *Proteobacteria*, *Actinobacteria*, and *Bacteroidetes*. Generators of genus spores such as *Clostridium* and *Bacillus* are very common, along with other gram-positive ones such as *Geodermatophilus* or *Streptococcus*. Also, *Alphaproteobacteria*, a very common bacterium class in soils (e.g., the family *Sphingomonadaceae*), are associated with dust [4, 9]. As regards *Archaea*, there are few studies of the atmosphere, in general, and of dust, in particular, that focus on this domain. Surely, reduced cases of pathogenic archaea have been studied to a lesser extent. *Aeropyrum* is the most detected genus of airborne archaea, but it is related to marine aerosols [11]. On the other hand, studies of pollen associated with dust are widespread. An interesting study investigated pollen transported from North Africa to Spain through Saharan dust and found that pollen from five non-native plant species was detected exclusively during dust events [99]. Lastly, viruses and virus-like particles have a great interest in the emission of dust. One study mentions virus-like particles associated with a transoceanic dust event. This report is based on epifluorescent microscopy of filters stained with a specific nucleic acid stain. An increase in the order of magnitude of virus-like particles was observed, from 104 to 2105 m^{-3} between the baseline condition and dust conditions in the Caribbean [41]. It is speculated that free airborne viruses show worse resistance to high ultraviolet radiation and dry air associated with long-distance transport in dust events resist worse than others [9].

8. Microbial diversity study in the atmosphere of the Iberian Peninsula after a Saharan dust intrusion

Four aerobiology sampling flights took place during February and March 2017 using the CASA C-212-200 aircraft from INTA. The study focused on microbial diversity in the atmosphere of the Iberian Peninsula during and after a Saharan dust intrusion. Flights took place under four different conditions: (1) during a strong Sahara dust storm that reached the north of the Iberian Peninsula, from February 22 to 24, 2017 (February 23, 2017) (**Figure** 7); (2) following precipitation (February 28, 2017); (3) following a dry period (March 8, 2017); and (4) along the northern coast of Spain (March 9, 2017). In each flight, samples were collected at different altitudes, and air samples were obtained simultaneously at ground level. A total of 20 samples were collected and are being analyzed. Cell presence was observed by scanning electron microscopy (SEM), and bacterial diversity is being studied by DNA extraction, 16S rRNA gene amplification, and Illumina MiSeq sequencing. Results are being analyzed via bioinformatics and biostatistical software (MOTHUR, SPSS, STAMP, CANOCO, and PAST) which will allow us to compare the results between the different flows and scenarios.

Although this study is not yet finished, some data can be advanced in this chapter. **Figure 6** shows SEM microphotographs obtained from samples in different scenarios. In general, the samples obtained during the days of dust intrusion (flight

Figure 7.
Saharan dust intrusion. Dust pours off the northweat Afrincan coast and blankets the Iberian Peninsula, 23 February, 2016. NASA satelital imagen via MODIS.

of February 23) appear completely covered with mineral particles. In these cases, more biological cells were detected than in the rest of the days. In the particular case of samples from the marine coast flight, more diatoms were observed (**Figure 6E**).

The analysis of diversity using the Shannon index showed that, in all cases, diversity was greater on days of Saharan dust intrusion, both in the samples taken from the ground and those taken at higher altitudes with the aircraft. This indicates that Saharan dust contributes microorganisms that are not present in the atmosphere on a daily basis. Diversity analysis showed phylum characteristics of soils, being *Alpha*- and *Betaproteobacteria* the most abundant classes. All of the analyses performed showed that bacterial diversity detected at ground level and in-flight samples during the dust intrusion event were similar among one another. The genus taxonomic levels of *Sphingomonas*, *Geodermatophilus*, *Methylobacter*, *Rhizobiales*, *Bacillus*, or *Clostridium* were present in every sample, but their sequences were more abundant in the case of ground samples and dust intrusion samples collected during the day flight. However, sequences of the genus *Flavobacterium*, *Streptococcus*, or *Cupriavidus* were most abundant in the case of samples collected during flight.

Preliminary conclusions show that bacterial diversity of airborne bacteria during days of dust intrusion is higher and similar to bacterial diversity commonly detected in soil samples. Further analyses are being conducted with these samples to obtain a complete description of the evolution of bacterial diversity during those days.

9. Conclusions

Intense UV radiation, low pressure, lack of water and nutrients, and freezing temperatures turn the atmosphere into an extreme environment, especially its upper layers. However, it is widely known that airborne bacteria, fungal spores, pollen, and other bioparticles exist. Numerous bacteria and fungi have been isolated and can survive even at stratospheric altitudes. Microbial survival in the atmosphere requires extremophilic characteristics, and therefore airborne microbiota is potentially useful for biotechnological applications. The role of airborne microbial communities is vital in the Earth, including interactions among the atmosphere, biosphere, climate, and public health. Airborne microorganisms are involved in meteorological processes and can serve as nuclei for cloud drops and ice crystals that precede precipitation, which influences the hydrological cycle and climate. Furthermore, their knowledge is essential in understanding the reproduction and propagation of organisms through various ecosystems. Furthermore, they can cause or improve human, animal, and plant diseases.

Airborne platforms that allow conducting a direct study of microorganisms in the atmosphere and molecular methodologies (e.g., "omics") could represent a major opportunity for approaching this question. Nevertheless, some challenges must yet be solved, such as low biomass, efficiency of sampling methods, the absence of standard protocols, or the high variability of the atmospheric environment.

Deserts and arid lands are one of the most important sources of aerosol emissions. Clouds of dust generated by storms mobilize tons of mineral particles, and it is known that microorganisms remain attached to the particles being transported over long distances. The large number of mineral particles and microorganisms thus placed into the atmosphere has global implications for climate, biochemical cycling, and health. North African soils, primarily the Sahara Desert, are one of the major sources of airborne dust on Earth. Saharan dust is often transported to southern Europe and could even reach high altitudes over the Atlantic Ocean and the European continent. Again, airborne platforms could be a perfect opportunity for conducting a direct study of the microbiology of this kind of events.

Acknowledgements

This work has been supported by grants from the Spanish government (www.mineco.gob.es and www.csic.es) (CGL2015-69758-P and i-LINK1151). Publishing fee has been has been supported by Open Access.

Author details

Ángeles Aguilera[1], Graciela de Diego-Castilla[1], Susana Osuna[1], Rafael Bardera[2], Suthyvann Sor Mendi[2], Yolanda Blanco[1] and Elena González-Toril[1]*

1 Astrobiology Institute (INTA-CSIC), National Institute for Aerospace Technology, Torrejón de Ardoz, Spain

2 Department of Aerodynamics and Propulsion, National Institute for Aerospace Technology, Torrejón de Ardoz, Spain

*Address all correspondence to: gonzalezte@cab.inta-csic.es

References

[1] Rothschild LJ, Mancinelli RL. Life in extreme environments. Nature. 2001;**409**:1092. DOI: 10.1038/35059215

[2] Smets W, Moretti S, Denys S, Lebeer S. Airborne bacteria in the atmosphere: Presence, purpose, and potential. Atmospheric Environment. 2016;**139**:214-221. DOI: 10.1016/j.atmosenv.2016.05.038

[3] Madigan MT, Marrs BL. Extremophiles. Scientific American. 1997;**276**(4):82-87

[4] Griffin DW. Atmospheric movement of microorganisms in clouds of desert dust and implications for human health. Clinical Microbiology Reviews. 2007;**20**(3):459-477. DOI: 10.1128/CMR.00039-06

[5] Smith D. Microbes in the upper atmosphere and unique opportunities for astrobiology research. Astrobiology. 2013;**13**:981-990. DOI: 10.1089/ast.2013.1074

[6] DasSarma P, DasSarma S. Survival of microbes in Earth's stratosphere. Current Opinion in Microbiology. 2018;**43**:24-30. DOI: 10.1016/J.MIB.2017.11.002

[7] Fröhlich-Nowoisky J, Kampf CJ, Weber B, Huffman JA, Pöhlker C, Andreae MO, et al. Bioaerosols in the earth system: Climate, health, and ecosystem interactions. Atmospheric Research. 2016;**182**:346-376. DOI: 10.1016/J.ATMOSRES.2016.07.018

[8] Griffin DW, Kellogg CA, Garrison VH, Lisle JT, Borden TC, Shinn EA. Atmospheric microbiology in the northern Caribbean during African dust events. Aerobiologia (Bologna). 2003;**19**(3):143-157. DOI: 10.1023/B:AERO.0000006530.32845.8d

[9] Kellogg CA, Griffin DW. Aerobiology and the global transport of desert dust. Trends in Ecology & Evolution. 2006;**21**:638-644. DOI: 10.1016/j.tree.2006.07.004

[10] Bowers RM, Lauber CL, Wiedinmyer C, Hamady M, Hallar AG, Fall RR, et al. Characterization of airborne microbial communities at a high-elevation site and their potential to act as atmospheric ice nuclei. Applied and Environmental Microbiology. 2009;**75**(15):5121-5130

[11] Smith DJ. Aeroplankton and the need for a global monitoring network. Bioscience. 2013;**63**(7):515-516. DOI: 10.1525/bio.2013.63.7.3

[12] Mayol E, Jiménez MA, Herndl GJ, Duarte CM, Arrieta JM. Resolving the abundance and air-sea fluxes of airborne microorganisms in the North Atlantic Ocean. Frontiers in Microbiology. 2014;**5**:557. DOI: 10.3389/fmicb.2014.00557

[13] Lindow SE, Brandl MT. Microbiology of the phyllosphere. Applied and Environmental Microbiology. 2003;**69**:1875-1883. DOI: 10.1007/BF02887579

[14] Amato P, Ménager M, Sancelme M, Laj P, Mailhot G, Delort AM. Microbial population in cloud water at the Puy de Dôme: Implications for the chemistry of clouds. Atmospheric Environment. 2005:4143-4153. DOI: 10.1016/j.atmosenv.2005.01.002

[15] Fierer N, Bradford MA, Jackson RB. Toward an ecological classification of soil bacteria. Ecology. 2007;**88**(6):1354-1364

[16] Fierer N, Liu Z, Rodríguez-Hernández M, Knight R, Henn M, Hernandez MT. Short-term temporal variability in airborne bacterial and fungal populations. Applied

and Environmental Microbiology. 2008;**74**(1):200-207. DOI: 10.1128/AEM.01467-07

[17] Burrows SM, Elbert W, Lawrence MG, Pöschl U. Bacteria in the global atmosphere—Part 1: Review and synthesis of literature data for different ecosystems. Atmospheric Chemistry and Physics. 2009;**9**(23):9263-9280. DOI: 10.5194/acp-9-9263-2009

[18] Burrows SM, Butler T, Jöckel P, Tost H, Kerkweg A, Pöschl U, et al. Bacteria in the global atmosphere—Part 2: Modeling of emissions and transport between different ecosystems. Atmospheric Chemistry and Physics. 2009;**9**(23):9281-9297. DOI: 10.5194/acp-9-9281-2009

[19] Smith DJ, Griffin DW, Jaffe DA. The high life: Transport of microbes in the atmosphere. Eos Transactions American Geophysical Union. 2011;**92**(30):249-250. DOI: 10.1029/2011EO300001

[20] Behzad H, Gojobori T, Mineta K. Challenges and opportunities of airborne metagenomics. Genome Biology and Evolution. 2015;7(5):1216-1226. DOI: 10.1093/gbe/evv064

[21] Griffin D. Non-spore forming eubacteria isolated at an altitude of 20,000 m in Earth's atmosphere: Extended incubation periods needed for culture-based assays. Aerobiologia. 2008;**24**:19-25. DOI: 10.1007/s10453-007-9078-7

[22] Smith DJ, Griffin DW, Schuerger AC. Stratospheric microbiology at 20 km over the Pacific Ocean. Aerobiologia (Bologna). 2010;**26**(1):35-46. DOI: 10.1007/s10453-009-9141-7

[23] Goudie AS, Middleton NJ. Desert dust in the global system. Springer Science & Business Media. 2006. pp. 1-11. DOI: 10.1007/3-540-32355-4

[24] Shao Y, Wyrwoll K-H, Chappell A, Huang J, Lin Z, GH MT, et al. Dust cycle: An emerging core theme in Earth system science. Aeolian Research. 2011;**2**:181-204. DOI: 10.1016/j.aeolia.2011.02.001

[25] Miller RL, Tegen I, Perlwitz J. Surface radiative forcing by soil dust aerosols and the hydrologic cycle. Journal of Geophysical Research—Atmospheres. 2004;**109**(D4):1-24. DOI: 10.1029/2003JD004085

[26] Song Z, Wang J, Wang S. Quantitative classification of northeast Asian dust events. Journal of Geophysical Research—Atmospheres. 2007;**112**(4):1-8. DOI: 10.1029/2006JD007048

[27] Ehrenberg CG. Neue Beobachlungen über blutartige Erscheinungen in Aegypten, Arabien und Sibirien, nebst einer Uebersicht und Kritik der früher bekannnten. Annalen der Physik. 1830;**94**(4):477-514. DOI: 10.1002/andp.18300940402

[28] Pasteur L. Expériences relatives aux générations dites spontanées. Comptes rendus hebdomadaires des séances de l'Académie des sciences. 1860;**50**:303-307

[29] Mora M, Perras A, Alekhova TA, Wink L, Krause R, Aleksandrova A, et al. Resilient microorganisms in dust samples of the international Space Station—Survival of the adaptation specialists. Microbiome. 2016;**4**:65. DOI: 10.1186/s40168-016-0217-7

[30] Despres VR, Huffman JA, Burrows SM, Hoose C, Safatov AS, Buryak G, et al. Primary biological aerosol particles in the atmosphere: A review. Tellus B. 2012;**64**:15598/1-15598/1559858. DOI: 10.3402/tellusb.v64i0.15598

[31] Smith DJ, Griffin DW, McPeters RD, Ward PD, Schuerger AC. Microbial survival in the stratosphere and implications for global dispersal. Aerobiologia (Bologna).

2011;**27**(4):319-332. DOI: 10.1007/s10453-011-9203-5

[32] Christner BC, Morris CE, Foreman CM, Cai R, Sands DC. Ubiquity of biological ice nucleators in snowfall. Science. 2008;**319**(5867):1214-1214. DOI: 10.1126/science.1149757

[33] Möhler O, Georgakopoulos DG, Morris CE, Benz S, Ebert V, Hunsmann S, et al. Heterogeneous ice nucleation activity of bacteria: New laboratory experiments at simulated cloud conditions. Biogeosciences. 2008;**5**(5):1425-1435. DOI: 10.5194/bg-5-1425-2008

[34] Morris CE, Conen F, Alex Huffman J, Phillips V, Pöschl U, Sands DC. Bioprecipitation: A feedback cycle linking Earth history, ecosystem dynamics and land use through biological ice nucleators in the atmosphere. Global Change Biology. 2014;**20**(2):341-351. DOI: 10.1111/gcb.12447

[35] Kim K-H, Kabir E, Kabir S. A review on the human health impact of airborne particulate matter. Environment International. 2015;**74**:136-143. DOI: 10.1016/J.ENVINT.2014.10.005

[36] Gonzalez-Martin C, Teigell-Perez N, Valladares B, Griffin DW. The global dispersion of pathogenic microorganisms by dust storms and its relevance to agriculture. Advances in Agronomy. 2014;**127**:1-41. DOI: 10.1016/B978-0-12-800131-8.00001-7

[37] Lighthart B. Mini-review of the concentration variations found in the alfresco atmospheric bacterial populations. Aerobiologia (Bologna). 2000;**16**(1):7-16. DOI: 10.1023/A:1007694618888

[38] Cho BC, Hwang CY. Prokaryotic abundance and 16S rRNA gene sequences detected in marine aerosols on the East Sea (Korea). FEMS Microbiology Ecology. 2011;**76**(2):327-341. DOI: 10.1111/j.1574-6941.2011.01053.x

[39] Smith DJ, Jaffe DA, Birmele MN, Griffin DW, Schuerger AC, Hee J, et al. Free tropospheric transport of microorganisms from Asia to North America. Microbial Ecology. 2012;**64**(4):973-985. DOI: 10.1007/s00248-012-0088-9

[40] DeLeon-Rodriguez N, Lathem TL, Rodriguez-R LM, Barazesh JM, Anderson BE, Beyersdorf AJ, et al. Microbiome of the upper troposphere: Species composition and prevalence, effects of tropical storms, and atmospheric implications. Proceedings of the National Academy of Sciences. 2013;**110**(7):2575-2580. DOI: 10.1073/pnas.1212089110

[41] Gonzalez-Martin C, Teigell-Perez N, Lyles M, Valladares B, Griffin DW. Epifluorescent direct counts of bacteria and viruses from topsoil of various desert dust storm regions. Research in Microbiology. 2013;**164**(1):17-21. DOI: 10.1016/j.resmic.2012.08.009

[42] Grinshpun SA, Buttner MP, Mainelis G, Willeke K. Sampling for airborne microorganisms. In: Manual of Environmental Microbiology. 4th ed. Washington, USA: American Society of Microbiology; 2016. pp. 2-3. DOI: 10.1128/9781555815882.ch74

[43] Hervàs A, Camarero L, Reche I, Casamayor E, Hervàs A, Camarero L, et al. Viability and potential for immigration of airborne bacteria from Africa that reach high mountain lakes in Europe. Environmental Microbiology. 2009;**11**:1612-1623. DOI: 10.1111/j.1462-2920.2009.01926.x

[44] Liu Y, Yao T, Jiao N, Kang S, Xu B, Zeng Y, et al. Bacterial diversity in

the snow over Tibetan plateau glaciers. Extremophiles: Life under Extreme Conditions. 2009;**13**:411-423. DOI: 10.1007/s00792-009-0227-5

[45] Hu W, Niu H, Murata K, Wu Z, Hu M, Kojima T, et al. Bacteria in atmospheric waters: Detection, characteristics and implications. Atmospheric Environment. 2018;**179**:201-221. DOI: 10.1016/J.ATMOSENV.2018.02.026

[46] Peccia J, Hernandez M. Photoreactivation in airborne mycobacterium parafortuitum. Applied and Environmental Microbiology. 2001;**67**(9):4225-4232. DOI: 10.1128/AEM.67.9.4225-4232.2001

[47] Lange JL, Thorne PS, Lynch N. Application of flow cytometry and fluorescent in situ hybridization for assessment of exposures to airborne bacteria. Applied and Environmental Microbiology. 1997;**63**(4):1557-1563

[48] Andersen AA. New sampler for the collection, sizing, and enumeration of viable airborne particles. Journal of Bacteriology. 1958;**76**(5):471-484

[49] Kesavan J, Sagripanti J-L. Evaluation criteria for bioaerosol samplers. Environmental Science: Processes & Impacts. 2015;**17**(3):638-645. DOI: 10.1039/C4EM00510D

[50] Haddrell AE, Thomas R. Aerobiology: Experimental considerations, observations, and future tools. Applied and Environmental Microbiology. 2017;**83**:1-15. DOI: 10.1128/AEM.00809-17

[51] Welam Henningson E, Lundquist M, Larsson E, Sandström G, Forsman M. A comparative study of different methods to determine the total number and the survival of bacteria in aerobiological samples. Journal of Aerosol Science. 1997;**28**:459-469. DOI: 10.1016/S0021-8502(96)00447-8

[52] Jensen PA, Todd WF, Davis GN, Scarpino PV. Evaluation of eight bioaerosol samplers challenged with aerosols of free bacteria. American Industrial Hygiene Association Journal. 1992;**53**(10):660-667. DOI: 10.1080/15298669291360319

[53] Terzieva S, Donnelly J, Ulevicius V, Grinshpun SA, Willeke K, Stelma GN, et al. Comparison of methods for detection and enumeration of airborne microorganisms collected by liquid impingement. Applied and Environmental Microbiology. 1996;**62**(7):2264-2272

[54] Ghosh B, Lal H, Srivastava A. Review of bioaerosols in indoor environment with special reference to sampling, analysis and control mechanisms. Environment International. 2015;**85**:254-272. DOI: 10.1016/J.ENVINT.2015.09.018

[55] Timmons DE, Fulton JD, Mitchell RB. Microorganisms of the upper atmosphere. Applied Microbiology. 1966;**14**(2):229-231

[56] Griffin DW, Kellogg CA. Dust storms and their impact on ocean and human health: Dust in Earth's atmosphere. EcoHealth. 2004;**1**:284-295. DOI: 10.1007/s10393-004-0120-8

[57] Yang Y, Itahashi S, Yokobori S, Yamagishi A. UV-resistant bacteria isolated from upper troposphere and lower stratosphere. Biological Sciences in Space. 2008;**22**(1):18-25. DOI: 10.2187/bss.22.18

[58] Kobayashi F, Maki T, Kakikawa M, Yamada M, Puspitasari F, Iwasaka Y. Bioprocess of Kosa bioaerosols: Effect of ultraviolet radiation on airborne bacteria within Kosa (Asian dust). Journal of Bioscience and Bioengineering. 2015;**119**(5):570-579. DOI: 10.1016/J.JBIOSC.2014.10.015

[59] Smith DJ, Griffin DW. Inadequate methods and questionable conclusions

in atmospheric life study. Proceedings of the National Academy of Sciences of the United States of America. 2013;**110**(23):E2084. DOI: 10.1073/pnas.1302612110

[60] Griffin DW. Terrestrial microorganisms at an altitude of 20,000 m in Earth's atmosphere. Aerobiologia (Bologna). 2004;**20**(2):135-140. DOI: 10.1023/B:AERO.0000032948.84077.12

[61] Pace NR. A molecular view of microbial diversity and the biosphere. Science. 1997;**276**:734-740. DOI: 10.1126/science.276.5313.734

[62] Brodie EL, DeSantis TZ, Parker JPM, Zubietta IX, Piceno YM, Andersen GL. Urban aerosols harbor diverse and dynamic bacterial populations. Proceedings of the National Academy of Sciences. 2007;**104**(1):299-304. DOI: 10.1073/PNAS.0608255104

[63] Hughes KA, McCartney HA, Lachlan-Cope TA, Pearce DA. A preliminary study of airborne microbial biodiversity over peninsular Antarctica. Cellular and Molecular Biology (Noisy-le-Grand, France). 2004;**50**(5):537-542. DOI: 10.1170/T543

[64] Kuske CR. Current and emerging technologies for the study of bacteria in the outdoor air. Current Opinion in Biotechnology. 2006;**17**(3):291-296. DOI: 10.1016/J.COPBIO.2006.04.001

[65] Maron P-A, Lejon DPH, Carvalho E, Bizet K, Lemanceau P, Ranjard L, et al. Assessing genetic structure and diversity of airborne bacterial communities by DNA fingerprinting and 16S rDNA clone library. Atmospheric Environment. 2005;**39**(20):3687-3695. DOI: 10.1016/J.ATMOSENV.2005.03.002

[66] Yoo K, Lee TK, Choi EJ, Yang J, Shukla SK, Hwang S, et al. Molecular approaches for the detection and monitoring of microbial communities in bioaerosols: A review. Journal of Environmental Sciences. 2017;**51**:234-247. DOI: 10.1016/J.JES.2016.07.002

[67] Bowers RM, McLetchie S, Knight R, Fierer N. Spatial variability in airborne bacterial communities across land-use types and their relationship to the bacterial communities of potential source environments. The ISME Journal. 2010;**5**:601. DOI: 10.1038/ismej.2010.167

[68] Franzetti A, Gandolfi I, Gaspari E, Ambrosini R, Bestetti G. Seasonal variability of bacteria in fine and coarse urban air particulate matter. Applied Microbiology and Biotechnology. 2011;**90**(2):745-753. DOI: 10.1007/s00253-010-3048-7

[69] Bertolini V, Gandolfi I, Ambrosini R, Bestetti G, Innocente E, Rampazzo G, et al. Temporal variability and effect of environmental variables on airborne bacterial communities in an urban area of Northern Italy. Applied Microbiology and Biotechnology. 2013;**97**(14):6561-6570. DOI: 10.1007/s00253-012-4450-0

[70] Gandolfi I, Bertolini V, Ambrosini R, Bestetti G, Franzetti A. Unravelling the bacterial diversity in the atmosphere. Applied Microbiology and Biotechnology. 2013;**97**(11):4727-4736. DOI: 10.1007/s00253-013-4901-2

[71] Rule AM, Kesavan J, Schwab KJ, Buckley TJ. Application of flow cytometry for the assessment of preservation and recovery efficiency of bioaerosol samplers spiked with *Pantoea agglomerans*. Environmental Science & Technology. 2007;**41**(7):2467-2472. DOI: 10.1021/es062394l

[72] Daims H, Brühl A, Amann R, Schleifer K-H, Wagner M. The domain-specific probe EUB338 is insufficient for the detection of all

bacteria: Development and evaluation of a more comprehensive probe set. Systematic and Applied Microbiology. 1999;**22**(3):434-444. DOI: 10.1016/S0723-2020(99)80053-8

[73] Amann RI. In situ identification of micro-organisms by whole cell hybridization with rRNA-targeted nucleic acid probes BT. In: Akkermans ADL, Van Elsas JD, De Bruijn FJ, editors. Molecular Microbial Ecology Manual. Dordrecht, Netherlands: Springer; 1995. pp. 331-345. DOI: 10.1007/978-94-011-0351-0_23

[74] Wallner G, Amann R, Beisker W. Optimizing fluorescent in situ hybridization with rRNA-targeted oligonucleotide probes for flow cytometric identification of microorganisms. Cytometry. 1993;**14**(2):136-143. DOI: 10.1002/cyto.990140205

[75] Delort A-M, Vaïtilingom M, Amato P, Sancelme M, Parazols M, Mailhot G, et al. A short overview of the microbial population in clouds: Potential roles in atmospheric chemistry and nucleation processes. Atmospheric Research. 2010;**98**(2-4):249-260. DOI: 10.1016/J.ATMOSRES.2010.07.004

[76] Bär M, Hardenberg J, Meron E, Provenzale A. Modelling the survival of bacteria in drylands: The advantage of being dormant. Proceedings of the Royal Society of London B: Biological Sciences. 2002;**269**(1494):937-942. DOI: 10.1098/rspb.2002.1958

[77] Tringe SG, Zhang T, Liu X, Yu Y, Lee WH, Yap J, et al. The airborne metagenome in an indoor urban environment. PLoS One. 2008;**3**(4):e1862. DOI: 10.1371/journal.pone.0001862

[78] Kennedy SP, Ng WV, Salzberg SL, Hood L, DasSarma S. Understanding the adaptation of Halobacterium species NRC-1 to its extreme environment through computational analysis of its genome sequence. Genome Research. 2001;**11**(10):1641-1650. DOI: 10.1101/gr.190201

[79] Ansari TU, Valsan AE, Ojha N, Ravikrishna R, Narasimhan B, Gunthe SS. Model simulations of fungal spore distribution over the Indian region. Atmospheric Environment. 2015;**122**:552-560. DOI: 10.1016/J.ATMOSENV.2015.10.020

[80] Heald CL, Spracklen DV. Atmospheric budget of primary biological aerosol particles from fungal spores. Geophysical Research Letters. 2009;**36**(9):L09806. DOI: 10.1029/2009GL037493

[81] Hummel M, Hoose C, Gallagher M, Healy DA, Huffman JA, O'Connor D, et al. Regional-scale simulations of fungal spore aerosols using an emission parameterization adapted to local measurements of fluorescent biological aerosol particles. Atmospheric Chemistry and Physics. 2015;**15**(11):6127-6146. DOI: 10.5194/acp-15-6127-2015

[82] Hoose C, Kristjánsson JE, Burrows SM. How important is biological ice nucleation in clouds on a global scale? Environmental Research Letters. 2010;**5**(2):024009. DOI: 10.1088/1748-9326/5/2/024009

[83] Sesartic A, Dallafior TN. Global fungal spore emissions, review and synthesis of literature data. Biogeosciences. 2011;**8**(5):1181-1192. DOI: 10.5194/bg-8-1181-2011

[84] Spracklen DV, Heald CL. The contribution of fungal spores and bacteria to regional and global aerosol number and ice nucleation immersion freezing rates. Atmospheric Chemistry and Physics. 2014;**14**(17):9051-9059. DOI: 10.5194/acp-14-9051-2014

[85] Middleton NJ. Desert dust hazards: A global review. Aeolian Research. 2017;**24**:53-63. DOI: 10.1016/J.AEOLIA.2016.12.001

[86] Acosta-Martínez V, Van Pelt S, Moore-Kucera J, Baddock MC, Zobeck TM. Microbiology of wind-eroded sediments: Current knowledge and future research directions. Aeolian Research. 2015;**18**:99-113. DOI: 10.1016/J.AEOLIA.2015.06.001

[87] Leys JF, Heidenreich SK, Strong CL, McTainsh GH, Quigley S. PM10 concentrations and mass transport during "red Dawn"—Sydney 23 September 2009. Aeolian Research. 2011;**3**(3):327-342. DOI: 10.1016/J.AEOLIA.2011.06.003

[88] Meola M, Lazzaro A, Zeyer J. Bacterial composition and survival on Sahara dust particles transported to the European Alps. Frontiers in Microbiology. 2015;**6**:1454. DOI: 10.3389/fmicb.2015.01454

[89] Peter H, Hörtnagl P, Reche I, Sommaruga R. Bacterial diversity and composition during rain events with and without Saharan dust influence reaching a high mountain lake in the Alps. Environmental Microbiology Reports. 2014;**6**(6):618-624. DOI: 10.1111/1758-2229.12175

[90] Griffin DW, Garrison VH, Herman JR, Shinn EA. African desert dust in the Caribbean atmosphere: Microbiology and public health. Aerobiologia (Bologna). 2001;**17**(3):203-213. DOI: 10.1023/A:1011868218901

[91] Moulin C, Lambert CE, Dulac F, Dayan U. Control of atmospheric export of dust from North Africa by the North Atlantic oscillation. Nature. 1997;**387**:691

[92] Perkins S. Dust, the thermostat how tiny airborne particles manipulate global climate. Science News. 2001;**160**(13):200-2002. DOI: 10.2307/4012776

[93] Rosselli R, Fiamma M, Deligios M, Pintus G, Pellizzaro G, Canu A, et al. Microbial immigration across the Mediterranean via airborne dust. Scientific Reports. 2015;**5**:16306

[94] Sánchez de la Campa A, García-Salamanca A, Solano J, de la Rosa J, Ramos J-L. Chemical and microbiological characterization of atmospheric particulate matter during an intense African dust event in southern Spain. Environmental Science & Technology. 2013;**47**(8):3630-3638. DOI: 10.1021/es3051235

[95] Itani GN, Smith CA. Dust rains deliver diverse assemblages of microorganisms to the Eastern Mediterranean. Scientific Reports. 2016;**6**:22657. DOI: 10.1038/srep22657

[96] Katra I, Arotsker L, Krasnov H, Zaritsky A, Kushmaro A, Ben-Dov E. Richness and diversity in dust stormborne biomes at the Southeast Mediterranean. Scientific Reports. 2014;**4**:5265. DOI: 10.1038/srep05265

[97] Karanasiou A, Moreno N, Moreno T, Viana M, de Leeuw F, Querol X. Health effects from Sahara dust episodes in Europe: Literature review and research gaps. Environment International. 2012;**47**:107-114. DOI: 10.1016/J.ENVINT.2012.06.012

[98] Díaz J, Linares C, Carmona R, Russo A, Ortiz C, Salvador P, et al. Saharan dust intrusions in Spain: Health impacts and associated synoptic conditions. Environmental Research. 2017;**156**:455-467. DOI: 10.1016/J.ENVRES.2017.03.047

[99] Cariñanos P, Galán C, Alcázar P, Domínguez E. Analysis of the particles transported with dust-clouds reaching Cordoba, Southwestern Spain. Archives of Environmental Contamination and Toxicology. 2004;**46**(2):141-146. DOI: 10.1007/s00244-003-2273-9

Chapter 4

Harnessing the Genetic Diversity and Metabolic Potential of Extremophilic Microorganisms through the Integration of Metagenomics and Single-Cell Genomics

Deepika Goyal, Shiv Swaroop and Janmejay Pandey

Abstract

Microorganisms thriving under extreme environments have proven to be an invaluable resource for metabolic products and processes. While studies carried out on microbial characterization of extremophilic environments during golden era of microbiology adapted a 'reductionist approach' and focused on isolation, purification and characterization of individual microbial isolates; the recent studies have implemented a holistic approach using both culture-dependent and culture-independent approaches for characterization of total microbial diversity of the extreme environments. Findings from these studies have unmistakably indicated that microbial diversity within extreme environments is much higher than anticipated. Consequently, unraveling the taxonomic and metabolic characteristics of microbial diversity in extreme environments has emerged as an imposing challenge in the field of microbiology and microbial biotechnology. To a great extent, this challenge has been addressed with inception and advancement of next-generation sequencing and computing methods for NGS data analyses. However, further it has been realized that in order to maximize the exploitation of genetic and metabolic diversity of extremophilic microbial diversity, the metagenomic approaches must be combined synergistically with single-cell genomics. A synergistic approach is expected to provide comprehensions into the biology of extremophilic microorganism, including their metabolic potential, molecular mechanisms of adaptations, unique genomic features including codon reassignments etc.

Keywords: extremophilic environments, metabolic diversity, metagenomics, single cell genomics, small molecule secondary metabolites

1. Introduction

There are a number of extreme ecosystems present on Earth that harbor an array of microorganisms with unique genetic diversity and metabolic capabilities [1, 2].

These unique capabilities enable them not only survive but also thrive in extremes of physicochemical parameters [3–6]. The idea that microorganisms might survive in such extreme environments and the term 'extremophile' was first proposed in the 1970s by Robert MacElroy. Conventionally, extremophilic microorganisms have been defined by their ability to grow optimally under environments characterized by extreme temperature, pH, pressure, and salinity etc. [7, 8]. It is argued that survival and growth under extreme environments require stabilization of cellular components and enzymes so that their optimal functionality is maintained. Therefore, extremophilic microorganisms are proposed to be one of the greatest reservoirs of the wide spectrum of exclusive enzymes and metabolites with significant biotechnological applications [9–15]. In addition, the extremophilic microorganisms are now also being regarded to have the pivotal role in maintaining the balance of global biogeochemical cycles [16–18]. With this understanding, there has been a continued increase in the scientific interest in isolation and characterization of extremophilic microorganisms. The same is clearly reflected by the fact that many new extremophilic microorganisms have been isolated and cultured in laboratories all over the world during the past 2–3 decades [19, 20]. Still, much of the physiological and phylogenetic diversity of extremophilic microorganisms remains rather unexplored. Given the ability of extremophilic microorganisms to thrive in the extreme environments; their taxonomic, genetic and metabolic characterization is widely regarded as an indispensable step towards harnessing their true potential. The progress in this line of scientific endeavor has remained hampered due to the vast majority of microbial biodiversity within extremophilic environments comprising of the lineages that are recalcitrant to traditional culturing techniques based isolation and purification approaches [21]. In absence of purified cultures of extremophilic microorganisms, the access to their genetic and metabolic diversity has remained obscure as only until recently, cultivability was the single most important prerequisite for having access to the genetic complement of individual organisms.

This limitation has been circumvented to a great extent with the implementation of culture-independent approach (i.e. metagenomics). The 'state of the art metagenomics technologies,' allow not only to develop a theoretical and mechanistic understanding of the possible role of extremophilic microorganisms in biogeochemical cycles but also assess the genetic & metabolic potentials (e.g. discover novel enzymes and proteins for industrial applications) of the uncultured extremophilic microbial population [22–25]. Having mentioned that, it is also pertinent to remark that even with the implementation of improved cultivation methodologies and metagenomics characterization, the understanding of the 'black box of extremophilic microbial diversity' has improved only marginally over the period of last 2 decades. The optimal exploitation of their potential still remains elusive. This situation could be attributed to the following reasons: (i) despite the ever-improving cultivation methodologies, most of the extremophilic microorganisms are not yet amenable to laboratory culturing which use traditional reductionist culturing approaches; (ii) the microbial biomass densities within extremophilic environments are often too less to yield enough DNA for carrying out effective culture-independent analyses (e.g. metagenomics, metatranscriptomics, and recombinant cloning of a gene of interest); and (iii) inability to annotate novel genetic complements during post-sequencing analyses of metagenomic due to lack of reference sequences in the nucleotide databases [24].

This situation demands continued improvement of technical methodologies towards assessing and harnessing the genetic and metabolic diversity of extremophilic microorganisms from even the minute quantities of retrievable metagenomic DNA. Some of the developments in this aspect have focused on improving the recovery of metagenomic DNA from extremophilic environments [26]. Yet another most important developments in this aspect has been the development of Single Cell Genome Analyses (SCGA) and its synergistic application with metagenomics [27]. The synergistic application of both of these approaches enables for assembly and annotation of draft genomes of even the uncultivated phyla. Therefore, these approaches could be effectively used to harness the genetic and metabolic potential of the extremophilic environments even without the need for extensive laboratory manipulation [28, 29]. Till date, such studies focusing on extreme environments have revealed substantial genomic information for several candidate extremophilic phyla, encompassing putative acidophiles, halophiles, thermophiles, and piezophiles. These data have also provided substantial insights (including catabolic and anabolic potential, molecular mechanism for adaptations to extreme environments, unique genomic features such as stop codon reassignments, and predictions about cell ultrastructure) into the biology of extremophilic microorganism. It is suggested that if metagenomics and SCGA methodologies are coupled with other "omics" technologies, such as transcriptomics, proteomics and metabolomics (i.e. study and quantification of mRNA transcript levels, proteins and cellular metabolites respectively), it could lead to further development of scientific capabilities for harnessing the genetic and metabolic potential of the extremophilic microbial diversity [30, 31].

2. Extremophilic environments and associated microbial diversity

The physicochemical characteristics of extremophilic environments observed on the planet Earth are quite diverse and they are often studied with regards to temperature, pH, salt concentration, nutrient availability etc. Some of the typical extremophilic environments widely studied include thermophilic environments, psychrophilic environments, halophilic environments, acidophilic environments, subterranean habitats, and hyper-arid environments [2, 8, 32]. Representative niches for each of these environments have been scanned with both cultivation-dependent and cultivation-independent approaches [19, 20, 24, 33]. A brief description of some of the representative extremophilic environments and the associated microbial diversity is presented below.

2.1 Thermophilic environments

Studies pertaining to thermophilic environments initiated in the 1970s and 1980s with the isolation of several novel hyperthermophiles. Subsequent studies led to the discovery of deep-sea hydrothermal vents and consequent addition of isolation of a wide range of hyper thermophilic microorganisms belonging to the 'archaeal' domain of the life [34]. During the 1990s, with the advent of culture-independent characterization of microbial diversity using 16S rRNA gene pool sequencing, the thermophilic environments e.g. hydrothermal vents were analyzed [35–37]. These studies could define the composition and diversity of the microbial communities present within the representative thermophilic environments and characterized the prokaryotic phylotypes amongst diverse thermophilic environments representing the temperature gradients from 60°C to 120°C [35–37]. However, the understanding of the functions associated with microbial diversity and the intra-species, inter-species interaction remained poorly defined.

A few of the culture-independent studies on thermophilic environments, which analyzed the sequence of the entire metagenomic DNA pool rather than just the phylogenetic marker gene, identified dominance of sulfur- recycling genes amongst the dominant phylotypes within the sulfur-rich deep-sea vents [35]. Similarly, the prevalence of hydrogen oxidation genes was observed in hydrogen-rich deep-sea hyperthermophilic vents [38–40]. A few other studies have identified the critical genetic signatures (e.g. genes for alternative mechanisms of nitrogen utilization) of the microbial communities surviving within the thermophilic environments. Some of the recent culture-independent studies on samples collected from thermophilic environments have indicated for the occurrence of the significantly higher diversity of CRISPR compared to the metagenomes of the mesophilic microbial diversity [41–43].

Even with increasing frequency of reports showing the identification of novel genetic and metabolic mechanisms prevalent in thermophilic environments; the comprehensive understanding about key genetic elements which determine the composition as well the function of the microbial diversity within the thermophilic environments is only poorly understood. It is not yet established how physicochemical factors contribute to shaping up the composition and structure of the microbial diversity of any thermophilic environment. The scenario is expected to improve only through the inclusion of physicochemical information along with full community metagenome data.

2.2 Psychrophilic environments

The psychrophilic environments are characterized by extremely low temperatures. Just like the thermophilic environments, they also represent one of the most thoroughly investigated extreme environments [21, 44]. It is noteworthy that unlike the thermophilic environments, the microbial diversity within psychrophilic environments consists of both eubacteria and archaea [45]. The biodiversity and adaptive strategies of psychrophilic microorganisms have been extensively studied. Results from some of the representative metagenomic studies on the psychrophilic environment have shown microbial community diversity and complexity to be significantly higher than other environments [45, 46]. The most note-worthy studies on psychrophilic environments have been carried out on samples from Antarctic continent, which harbors sub-glacial ice habitat. These studies have reported the dominance of 'chemoautotrophs' that are capable of tapping reduced iron and reduced sulfur compounds as the source of energy [47]. Other studies with psychrophilic environments have recognized the presence of 'chemolithotrophic' bacterial and archaeal communities [45, 47]. These share a close phylogenetic relationship with microorganisms able to use reduced nitrogen, and iron compounds as the source of energy. With regards to the psychrophilic environments, it is generally accepted that 'availability of organic metabolizable carbon' is the single most dominant factor determining the microbial activity, diversity, and dynamics.

2.3 Acidophilic environments

Acidophilic environments have emerged as 'extremophilic environments of choice' for studies on mechanisms and genetic elements determining the survival of life under extreme environments. A number of studies had reported attempts for isolation of microorganisms from acidophilic environments. Culture-independent studies with respect to acidophilic environments were first carreid out with a natural acidophilic biofilm sample [48]. Subsequent studies in this regard were carried out on samples collected from an Acid Mine Drainage located at different parts of the world [49–52]. The data obtained with these samples showed the microbial community structure to have a poor diversity with presence of only chemoautotrophic consortia largely comprising members of genera *Leptospirillum* and *Ferroplasm* [48]. The genetic signatures observed within the Acid Mine Drainage metagenomes indicated for molecular mechanisms for acidophilic survival through implmentation of unique carbon metabolic pathways for Carbon metabolism, Nitrogen fixation and iron oxidation [53]. The community composition of Acid Mine Drainage samples were found to have significant contrast to the naturally occuring acidophilic biofilms that has *Acidithiobacillus*, *Acidimicrobium* and *Ferrimicrobium* as the dominant genera present within the community [49, 52, 54]. The other noticeably dominant microbial extremophilic taxa in acid mine drainages was Ferroplasm and Thermoplasmatales archaea [55, 56].

2.4 Halophilic environments

Like other extremophilic environments, the microbial community structure and diversity of the halophilic environments has also been subject of great scientific curiosity. Several culture dependent and culture independent studies have been carried in past 2–3 decades for the assessment of the microbial diversity thriving within the halophilic environments [57–59]. The research findings from some of the most important studies have been thoroughly reviewed. Studies pertaining to halophilic microorganisms have got greatly benefitted with the implementation of cultivation independent approaches for microbial diversity analyses. Metagenomic analyses of the samples collected from multiple hypersaline systems (e.g. Tyrell Lake, Crystallizer Ponds) have indicated presence of high phylotypic diversity with the dominance of halophilic archaeon in particular [60–63]. The whole DNA pool metagenome sequencing of halophilic samples followed by *de novo* assembly and annotation resulted in discovery of a dominant novel uncultivated archaeal class viz., Nanohaloarchaea [60]. This study also revealed occurrence of a unique combination of amino acids which increase the structural flexibility and osmo-resistance of the protein elements. Another characteristic feature of the genetic resources associated with microbial diversity within halophilic environment was discovered in an independent study and it was observed to be the prevalence of Halo-resistance mechanisms orchestrated through synthesis of solutes (such as glycine, betaine, ectoine and trehalose etc.) that are compatible with high salt concentrations [64].

3. Extremophilic microorganisms: invaluable source of novel metabolites

Microorganisms surviving in the extreme environments are being looked up to as they could help treat a wide spectrum of human illnesses, from ovarian cancer, migraine, high blood pressure, ovarian cancer and lung cancer to Alzheimer's disease. This doctrine has emerged out of the understanding that extremophilic environments present very hostile conditions that impose serious threat to survival of any organism exposed to them [8, 65]. However, extremophilic microorganisms which thrive under such hostile environment must be doing it by synthesizing unusual, but potentially very useful, secondary metabolites. Probably, the best studied molecules produced by extremophilic microorganisms are (i) biocatalytic proteins that are often referred as extremozymes; and (ii) secondary metabolites that are not directly required growth of the microorganism, yet they often perform many helpful functions, such as enabling defense mechanisms etc. [66–69],

It is suggested that extremophile enzymes would be more suitable and stable for use in industrial biotechnology applications than those obtained from mesophilic microbial species [9, 21, 70]. Also, the unusual secondary metabolites isolated from extremophilic microorganisms are steadily being characterized as drug molecules with unique potential and applications. One of the recently published studies reported characterization of a secondary metabolite (viz., dihydrogranaticin) from a thermophilic fungus exhibits wide spectrum antibiotic functions. Similarly, secondary metabolites isolated from a psychrophilic bacterium from the Arctic glaciers have been reported to inhibit the growth of human colon cancer cells. Another secondary metabolite (psychrophilin D) isolated from a psychrophilic microorganism, exhibits inhibitory activity against mouse leukemia cell line [71].

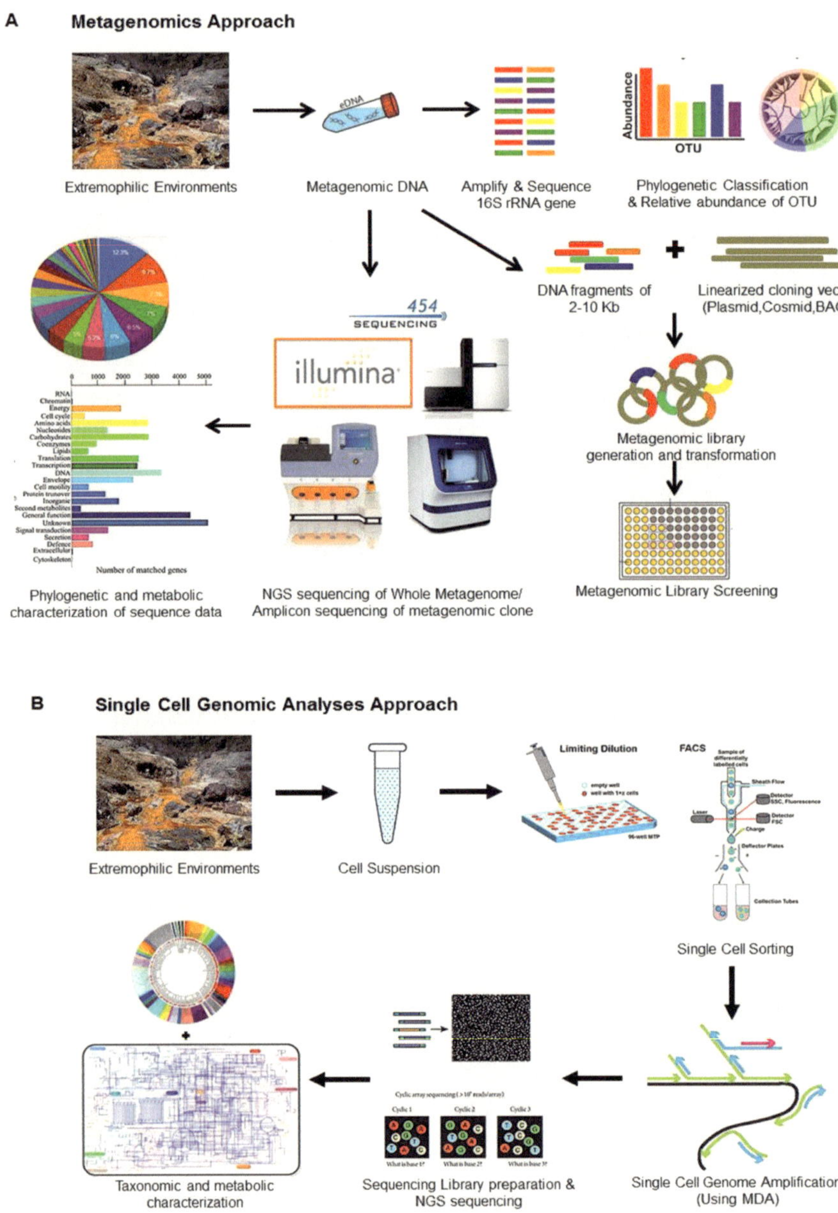

Figure 1.
Schematic representation of the workflow used for culture independent approaches for characterization of microbial diversity viz., metagenomics (A) and single cell genomics (B).

In accordance to the other extremophilic environments, the secondary metabolites produced by organisms that thrive at acidophilic environments, are a valuable source of novel metabolites. According to a recent survey, more than 20 previously unknown natural products have been isolated from acidophilic microbial diversity. Another valuable type of metabolites that is being proposed to have significant technological application is the 'natural inhibitors' of therapeutic target proteins. A representative example of such natural inhibitors was reported as berkeleyamide A, a secondary metabolite isolated from acidophilic strains *Penicillium rubrum* species [72]. This inhibits the proteases caspase-1 and matrix metalloproteinase 3 (MMP3), both of which are implicated in malignancy of some of the cancer types [72]. Another molecule (i.e. Berkelic acid,) isolated from an extremophilic microorganism, has a very unusual tetracyclic structure and it also inhibits both caspase-1 and MMP-3 [73]. Consequently, it exhibits selective inhibitory activity against an ovarian cancer cell line which has implication of abovementioned genes in cancer progress. Unfortunately, there is significantly less information available relevant to the secondary metabolites produced by extremophilic microorganism thriving at high pH and high salt concentrations. It has been often suggested that the enzymes from these microorganisms would be quite useful a biological detergents.

Considering the well-established potentials of the metabolites of the extremophilic origin, there is a need to develop fundamental understanding with respect to their physiological role in the growth and survival of extremophilic microorganism as well as their adaptation to the hostile environment. Many of the metabolites remain 'cryptic' during the cultivation of the extremophiles under the *in vitro* conditions since recreating the physicochemical conditions observed in the extreme environments within the laboratory is technically challenging, complicated and expensive [74]. Metagenomic Analyses and Single Cell Genomic Analyses., which enable the assessment of genetic and metabolic diversity without the need of cultivating the microorganisms, have helped to circumvent the limitations caused by the cryptic nature of secondary metabolic genes [75–77]. **Figure 1** presents a schematic representation of the workflow used for the metagenomics (**Figure 1A**) and single cell genomics (**Figure 1B**). As of now, a number of studies have already been carried out with metagenomes and single cell genomes from the extreme environments for studying extremozymes and cryptic metabolites.

4. Cultivation-independent approaches: tapping extremophilic metabolites

Cultivation-independent approaches are based on direct isolation of whole metagenomic DNA/environmental DNA. Subsequent downstream treatments of metagenomic DNA are broadly classified into 2 categories, i.e. (i) Metagenomic library generation and its functional screening; and (ii) Direct sequencing of the whole metagenomic DNA content [78–82]. The same technological framework is applicable to metagenomic analyses of samples collected from extremophilic environments. However, the complex nature of extremophilic matrix presents certain unique technical challenges with respect to isolation of metagenomic DNA. The methodologies successfully implemented to mesophilic sites for metagenomic DNA isolation often tend to be non-sufficient for isolation of high quality and high quantity metagenomic DNA from extremophilic samples. Even with non- optimal metagenomic DNA isolation procedures, the cultivation independent approach has enabled identification and characterization of several valuable extremozymes and extremophilic metabolites [20, 75].

4.1 Functional screening of extremophilic metagenomic libraries

Using rather simple and direct readout assays (e.g. appearance of either a halo or a color), the functional screening of the metagenomic libraries have been carried out for a number of extremophilic environments. For example, in a recent study, the Antarctic desert soil metagenomic library was screened for psychrophilic esterases using agar plates based screening approach [83, 84]. The positive clone with desired activity was selected on the basis of formation of a clear halo around the metagenomic clone. The halo formation indicated tributyrin hydrolysis; and resulted in identification and characterization of a novel cold-active psychrophilic esterase. Noticeably, it was found to be only distantly related to previously reported lipases.

While the abovementioned example for isolation and characterization of a novel psychrophilic esterase clearly highlights the value of 'functional screening' of the metagenomic libraries of the extremophilic origin, yet, it is also well acknowledged that many of the extremozymes and extremophilic metabolites are not expressed from the clones of the metagenomics library and therefore, they are not amenable to identification by library screening assays [85]. Several attempts have been made to evade the apparent limitations associated with library screening approach to metagenomics. Screening and development of alternative host for functional metagenomics screening [86] and development and application of 'Reporter Vectors' has been one of the most distinct attempts in this regard. One such Reporter Vector for metagenomic library screen was developed to have product-induced gene-expression of a reporter gene. It is done by coupling of the reporter gene to a product-sensitive transcription factor; Thus upon formation of a desired product, the transcription of the reporter gene is initiated, which could be subsequently monitored through standard reporter gene assay (e.g. fluorescence) [87]. Complementation assays have also been used as a strategy for functional screening and isolation of novel biocatalysts from metagenomic libraries [88].

4.2 Homology search based screening of the extremophilic metagenome

The alternative approach is based on direct sequencing and homology search based screening of the gene(s), protein(s) and secondary metabolites of interest. This approach generally involves DNA amplification (PCR) step as a necessary step of sequencing procedures. Even the next generation sequencing platforms (i.e. Pyrosequencing, Sequencing by Synthesis, and Ion Sequencing) involve the step for PCR amplification of the metagenomic Pool DNA [20, 53, 79]. An earlier approach for homology search based screening of metagenomic DNA used 'heterologous probe –hybridization'; however, that approach has given way to NGS approaches. With advancement in the field of genome informatics, and metagenome informatics, it is now easily feasible to detect conserved enzymatic sequence motifs in metagenomic DNA sequences including the metagenomes of extremophilic environments [24]. The most noticeable advantage of this approach over the functional screening of the metagenomic library is the inherent high throughput and flexibility to extend the scope of screening using *in silico* homology search and screening [82]. The sequence homology search based screening approach has been used with primary metagenomic sequence data as well as with the pre- existing metagenomic datasets. The homology search based screening of metagenomic sequences gets limited only in terms of the 'existing sequence databases'. In other words any novel sequence(s) with significant divergence from the previously characterized/reference sequences or not having homology gets identified as "sequence with unknown function".

5. Diversity of products screened from extremophilic metagenomes

In addition to the screening for extremozymes, the extremophilic metagenomes have also been subjected to screening for identification of small molecules and secondary metabolites which could have potential pharmaceutical applications as antibiotics, antifungal, anti-inflammatory, anti-tuberculosis, anti-cancer and immunosuppressive etc. Both functional screening of metagenome library and homology search based screening approaches have been successfully used for this purpose [22, 24, 36, 40, 62]. In comparison the extremozymes, there are relatively fewer high throughput assays available for detecting metagenomic clones that can produce small molecules and/or secondary metabolites. Thus functional screening has not been used very often for metagenomic libraries with the objective of identifying novel secondary metabolites. Therefore, there is a constant need for development of innovative functional screening methods for identification of small molecules and secondary metabolites of extremophilic origin. A few discreet studies have shown examples of novel screening approaches. In one such example a novel screening method was developed with use of indicator "Chrome Azurol-S" (CAS), which undergoes chromogenic change from orange to blue in the presence of iron. This screening method was subsequently used for identification of metagenomic clones (as well as cultivable isolates) encoding siderophores (the iron chelators). In these studies gene clusters encoding novel siderophores were identified from novel uncultivable strains.

In comparison to the functional screening, the homology search screening has been more frequently used for screening of metagenomes for the extremophilic metabolites. For the homology search screening, the metagenomic sequence data is probed to identify gene(s)/gene cluster(s) containing conserved domains or sequence that are predicted to be associated with biosynthesis of a secondary metabolites of interest. The most prominent secondary metabolites identified through homology search screening of extremophilic metagenome sequences has successfully led to the identified and characterization of: (i) glycopeptide antibiotics; (ii) cyanobactins cytotoxins; (iii) type –II polyketides antibiotics and anticancer molecules; and (iv) Trans-acyltransferse (trans-AT) polyketides [89–93]. While each of these classes of small molecules/secondary metabolites have been previously identified and characterized from the cultivable microbial diversity (more specifically actinobacterial diversity), however, with use of homology search screening of the extremophilic metagenomes, a number of novel representatives of the chemical scaffolds have been successfully identified and characterized.

5.1 Identification and characterization of glycopeptide

Glycopeptides are small molecule secondary metabolites produced by diverse organisms ranging from Proteobacteria to higher plants with Actinobacteria being the single most important source. These small molecules exhibit antibacterial activity against some of the most resistant Gram-positive pathogenic bacteria [94]. Consequently, glycopeptide are molecules of great scientific and industrial significance. The assortment of glycopeptides isolated and characterized from cultivable bacterial diversity is only very limited; therefore, several studies have been carried out with the objective of widening the catalogue of the glycopeptides through exploitation of culture- independent approaches. In one such study, soil metagenome was used as the DNA template and used for amplification a gene corresponding to OxyC, an oxidation coupling enzyme which is highly conserved and catalyzes a vital intermediate reaction during synthesis of many glycopeptides. This approach resulted in identification of multiple predicted glycopeptide-encoding gene clusters from the soil metagenomic libraries. In the follow up studies, the novel glycopeptide synthesis related gene(s) and gene cluster(s) identified from the metagenomic DNA were transformed and heterologously expressed in a *Streptomyces* expression host [95, 96]. Such technical intervention resulted in several new derivative glycopeptide antibiotics (with methyl, sulfur and sugar substitution) were generated being synthesized.

5.2 Identification and characterization of cyanobactins

Cyanobactins are a family of small, cyclic peptides produced by cyanobacteria and consist of N-to-C macro-cylization of a 6–20 amino acid chain. They are generally assembled through the cleavage and modification of short precursor proteins. Many of these peptides show antimalarial or antitumor activity [97]. It is speculated that close to 30% of all cyanobacterial strains contain genes corresponding to synthesis of cyanobactins [98, 99]. It is also speculated that, bacterial diversity other than cyanobacteria may also have harbor the gene(s) and gene cluster(s) for synthesis of cyanobactins [98]. However, access to such cyanobactins gene cluster(s) is limited due to the non- cultivability of the vast microbial majority. A few metagenomic studies have reported cloning and heterologous expression of biosynthetic gene clusters for the cyanobactins. In one such example study, the gene cluster for 'patellamide' was cloned and heterologously expressed from metagenomic libraries of uncultured cyanobacterial symbionts associated with marine sponge [100, 101]. In other studies, the structural diversity of diversity was enriched with subtle changes in the gene encoding for precursor peptide and employed it in combination with multiple strategies e.g. (i) orthogonal loading of unnatural amino acids; (ii) mutagenesis of precursor peptide; (iii) generation of a library of hybrid cyanobactins [90].

5.3 Identification and characterization of Type II polyketides

Type II polyketides are a group of small molecules with aromatic rings and contain alternating carbonyl and methylene groups (-CO-CH2-).Many of the Type II polyketides (e.g. tetracycline and doxorubicin) are well documented for antimicrobial and ant cancerous activities [90]. Gene clusters involved in synthesis of these small molecules are rather divergent and exhibit low levels of DNA sequence homology, yet each of them contain at least a 'polyketide synthetase', encoded by three highly conserved genes, i.e. 2 genes for ketosynthases (KSs) and one gene for a acyl carrier protein. These 3 genes are referred as 'minimal PKS synthesis gene cluster'. Studies carried out with metagenomes in general and extremophilic metagenome in particular have shown a rich diversity of novel 'minimal PKS synthesis gene cluster' [102]. In subsequent studies, gene clusters with minimal PKS synthesis genes were identified in soil metagenomes [103]. The transformation and heterologous expression in different strains belonging to genus *Streptomyces* and lead to synthesis and identification of several new polyketide metabolites with previously unknown and rare carbon skeletons [93].

5.4 Identification and characterization of trans-acyltransferse polyketides

This class of small molecule polyketides is biosynthesized through activity of a freestanding acyltransferases and constitutes one of the most important groups of pharmacologically interesting polyketides. Considering their pharmaceutical implication and rather limited catalogue from the cultivated microorganism, the metagenomic route of discovery has been adapted. In this approach, the metagenomes from various environments including the extremophilic environments have been probed for presence of a conserved trans-Acyltransferase specific DNA sequences [104]. Using this approach, a single amplicons have been identified which would produce the novel Trans-acyltransferse polyketides. Unlike the Type II polyketides, studies with heterologous transformation and expression of the Trans-acyltransferse polyketides are relative obscure, yet, a few discreet studies have shown genesis of hybrid Trans-acyltransferse polyketides. In one such examples study, a gene encoding for O-methyltransferase from the pederin gene clusters was transformed in a mycalamide-A producing strains. Upon expression the O-methyltransferase catalyzed a site-specifically methylation which resulted in production of a hybrid compound 18-O-methylmycalamide which showed significantly improved antitumor activities [105].

6. Single Cell Genome Analyses of the extremophilic microbial diversity

A recent concept in the field of the culture- independent approaches for identification and characterization of microbial genetic and metabolic diversity is "Single Cell Genome Analyses (SCGA)" [106]. This approach accesses genomes from one cell at a time. Therefore, this approach allows the analyses of the microbial genetic and metabolic diversity at the level of the most fundamental biological unit. The central technical aspect of this approach involves separation of individual cells from a complex mixture of environmental matrix using a cell sorting methods such as fluorescence-activated cell sorting (FACS). Cell separation is followed by cell lysis and recovery of the femtogram levels of DNA from a Single cell. The recovered single cell DNA is amplified using multiple displacement amplification (MDA) and amplification of single cell genomic DNA, such that the quantities of DNA increases to 100s of nano grams – 10s of micro grams (a 10^3-10^6 fold increase) [107, 108]. The single amplified genomes (SAGs) are subsequently used for screening by PCR amplification and NGS sequencing. The taxonomic identity of the concerned extremophilic microbial cell is ascertained with 16S rRNA gene sequencing, whereas subsequent shotgun or NGS sequencing, assembly and annotation is carried out with single amplified genomes of interest identified through preliminary phylotype characterization [106–109].

Despite its tremendous scientific capabilities, the SCGA is yet to make outreaching impact on microbial genomics in general and extremophilic microbiology in particular. The technical procedure used for SCGA faces many challenges that are not yet completely addressed. The most critical challenges include: (i) technical limitation in precise and reproducible separation of single bacterial cells with available methodologies; (ii) low amounts of starting DNA recoverable from single bacterial cell; (iii) requirement of a high degree of amplification; (iv) possibility of cross contamination; (v) introduction of chimeric artifacts and biases in genomic coverage during single genome amplification; and (v) poor post-sequencing quality control, data analyses and sequence assembly [110]. Due to these limitations, the resulting composite assemblies from SCGA can often represent incomplete or inaccurately characterized genomes for a given strain or species [107, 111]. However, several technological updates are being made to circumvent these limitations of the SCGA, which would soon enable highly accurate data generation and its physiological interpretation based on the absence as well as presence of genes and pathways [108].

6.1 Combining single cell genomics and metagenomics

Despite the individual technical limitations of both the approaches, it is regarded that the combined synergistic application of single-cell genomics and metagenomics can offer great opportunities, since the advantages offered by each of these techniques are complementary in nature. To highlight, it is underlined that one hand metagenomics is not known to suffer from any problem associated with chimera generation during strand displacement and genome amplification or separation of individual microbial cells from a complex heterogeneous mixture. On the other hand single-cell genomics overcomes the limitation of metagenomics by leading to a direct and unambiguous association of phylogeny and metabolic functions. Information obtained from SCGA can be effectively used to assign taxonomy to individual metagenome contigs with high accuracy [107, 112–114]. SCGA may also be used for retrieving complete genomes of candidate taxon from the metagenomic data. Similarly, the metagenomic reads can be mapped back to scaffolds for closely related SAG and therefore significantly improve their annotation.

The synergistic application of metagenomics and single cell genomics is regarded to have a unified and far reaching implication in harnessing the biotechnological potential of the extremophilic microbial diversity. As a matter of fact, extremophilic environments have already featured prominently in studies implementing both metagenomics and single-cell genomics studies. The most note-worthy set of studies were performed on acidophilic biofilms of Richmond Mine, California, USA, wherein initial metagenomic studies led to the identification of dominant microbial communities, while subsequent single cell genomics studies could identify even novel, low-abundance archaeal lineages that were later named as archaeal richmond mine acidophilic nanoorganisms (ARMAN) [115, 116]. The nanoorganisms have since been the matter of investigation throughout the world. In the same vein, the synergistic application of metagenomics and single cell genomics has led to identification of three previously uncultivated and uncharacterized halophilic phylotypes that represent the candidate phylum Nanohaloarchaeota from studies carried out on samples collected from halophilic Pola salterns, Alicante, Spain. Apart from the taxonomic and phylogenetic characterization of novel extremophiles, the synergistic application of metagenomics and single cell genomics also led to identification of their critical metabolic functions e.g. presence of rhodopsin and genes for a photoheterotrophic lifestyle.

7. Conclusion

The advent of 'culture independent' approaches for characterization of microbial diversity and their dynamics has been the single most significant development in the field of microbiology in general and microbial ecology, microbial biotechnology in particular. It has also greatly accelerated the research pertaining to extremophilic microbial diversity. With use of present 'state of the art' technologies viz., metagenomics and single cell genomics, a number of vital discoveries have been made that would not have been possible without the use of these technologies. Thus, it could be proposed that although, considerable progress has been made, yet there is a lot of scope for better application of metagenomics and single-cell genomics approaches to not only access genomes for discovering novel taxonomic lineages of extremophilic microorganisms but also harness their genetic and metabolic potential towards discovery of novel high value metabolites.

As an eventual future objective, the application of metagenomics and single cell genomics would be expected to complement the traditional cultivation approaches and follow suit with 'genomics guided –microbial culturing' towards' establishment of knowledge of the metabolic interactions circuits within mesophilic environments and more specifically within the extremophilic environments. Exploration of cultivation- independent approaches promise an exciting future for assessment and exploitation of extremophilic microbial diversity.

Conflict of interest

Authors declare 'no conflict of interest' with respect to publication of this book chapter.

Author details

Deepika Goyal[1†], Shiv Swaroop[2†] and Janmejay Pandey[1*]

1 Department of Biotechnology, School of Life Sciences, C.U. Rajasthan, India

2 Department of Biochemistry, School of Life Sciences, C.U. Rajasthan, India

*Address all correspondence to: janmejay@curaj.ac.in

† Both the authors contributed equally.

References

[1] Hu X. Ciliates in extreme environments. The Journal of eukaryotic microbiology. 2014;**61**:410-418

[2] Harrison JP, Gheeraert N, Tsigelnitskiy D, Cockell CS. The limits for life under multiple extremes. Trends in microbiology. 2013;**21**:204-212

[3] Burtscher M, Gatterer H, Burtscher J, Mairbaurl H. Extreme Terrestrial Environments: Life in Thermal Stress and Hypoxia. A Narrative Review. Frontiers in physiology. 2018;**9**:572

[4] Bang C, Dagan T, Deines P, Dubilier N, Duschl WJ, Fraune S, et al. Metaorganisms in extreme environments: do microbes play a role in organismal adaptation? Zoology (Jena, Germany). 2018;**127**:1-19

[5] Chevin LM, Hoffmann AA. Evolution of phenotypic plasticity in extreme environments. Philosophical transactions of the Royal Society of London. Series B, Biological sciences. 2017;**372**

[6] Morozkina EV, Slutskaia ES, Fedorova TV, Tugai TI, Golubeva LI, Koroleva OV. Extremophilic microorganisms: biochemical adaptation and biotechnological application (review). Prikladnaia biokhimiia i mikrobiologiia. 2010;**46**:5-20

[7] Stetter KO. Extremophiles and their adaptation to hot environments. FEBS letters. 1999;**452**:22-25

[8] Rothschild LJ, Mancinelli RL. Life in extreme environments. Nature. 2001;**409**:1092-1101

[9] Littlechild JA. Enzymes from Extreme Environments and Their Industrial Applications. Frontiers in bioengineering and biotechnology. 2015;**3**:161

[10] Antranikian G, Vorgias CE, Bertoldo C. Extreme environments as a resource for microorganisms and novel biocatalysts. Advances in biochemical engineering/biotechnology. 2005;**96**:219-262

[11] van den Burg B. Extremophiles as a source for novel enzymes. Current opinion in microbiology. 2003;**6**:213-218

[12] Elleuche S, Schroder C, Sahm K, Antranikian G. Extremozymes--biocatalysts with unique properties from extremophilic microorganisms. Current opinion in biotechnology. 2014;**29**:116-123

[13] Cavicchioli R, Siddiqui KS, Andrews D, Sowers KR. Low-temperature extremophiles and their applications. Current opinion in biotechnology. 2002;**13**:253-261

[14] Ferrer M, Golyshina O, Beloqui A, Golyshin PN. Mining enzymes from extreme environments. Current opinion in microbiology. 2007;**10**:207-214

[15] Schiraldi C, De Rosa M. The production of biocatalysts and biomolecules from extremophiles. Trends in biotechnology. 2002;**20**:515-521

[16] Sorokin DY, Berben T, Melton ED, Overmars L, Vavourakis CD, Muyzer G. Microbial diversity and biogeochemical cycling in soda lakes. Extremophiles: life under extreme conditions. 2014;**18**:791-809

[17] Offre P, Spang A, Schleper C. Archaea in biogeochemical cycles. Annual review of microbiology. 2013;**67**:437-457

[18] Bao P, Li GX, Sun GX, Xu YY, Meharg AA, Zhu YG. The role of sulfate-reducing prokaryotes in the coupling of element biogeochemical cycling. The Science of the total environment. 2018;**613-614**:398-408

[19] Rinker KD, Han CJ, Kelly RM. Continuous culture as a tool for investigating the growth physiology of heterotrophic hyperthermophiles and extreme thermoacidophiles. Journal of applied microbiology. 1998;**85**(Suppl 1): 118s-127s

[20] Vester JK, Glaring MA, Stougaard P. Improved cultivation and metagenomics as new tools for bioprospecting in cold environments. Extremophiles: life under extreme conditions. 2015;**19**:17-29

[21] Feller G. Life at low temperatures: is disorder the driving force? Extremophiles: life under extreme conditions. 2007;**11**:211-216

[22] Saxena R, Dhakan DB, Mittal P, Waiker P, Chowdhury A, Ghatak A, et al. Metagenomic Analysis of Hot Springs in Central India Reveals Hydrocarbon Degrading Thermophiles and Pathways Essential for Survival in Extreme Environments. Frontiers in microbiology. 2016;**7**:2123

[23] Vavourakis CD, Ghai R, Rodriguez-Valera F, Sorokin DY, Tringe SG, Hugenholtz P, et al. Metagenomic Insights into the Uncultured Diversity and Physiology of Microbes in Four Hypersaline Soda Lake Brines. Frontiers in microbiology. 2016;**7**:211

[24] Cowan DA, Ramond JB, Makhalanyane TP, De Maayer P. Metagenomics of extreme environments. Current opinion in microbiology. 2015;**25**:97-102

[25] Poli A, Finore I, Romano I, Gioiello A, Lama L, Nicolaus B. Microbial Diversity in Extreme Marine Habitats and Their Biomolecules. Microbiome. 2017;**5**

[26] Verma D, Satyanarayana T. An improved protocol for DNA extraction from alkaline soil and sediment samples for constructing metagenomic libraries. Applied biochemistry and biotechnology. 2011;**165**:454-464

[27] Chen Z, Chen L, Zhang W. Tools for Genomic and Transcriptomic Analysis of Microbes at Single-Cell Level. Frontiers in microbiology. 2017;**8**:1831

[28] Lutz S, Anesio AM, Field K, Benning LG. Integrated 'Omics', Targeted Metabolite and Single-cell Analyses of Arctic Snow Algae Functionality and Adaptability. Frontiers in microbiology. 2015;**6**:1323

[29] Zhang W, Li F, Nie L. Integrating multiple 'omics' analysis for microbial biology: application and methodologies. Microbiology (Reading, England). 2010;**156**:287-301

[30] Y. Seeleuthner, S. Mondy, Single-cell genomics of multiple uncultured stramenopiles reveals underestimated functional diversity across oceans, 9 (2018) 310.

[31] Kodzius R, Gojobori T. Single-cell technologies in environmental omics. Gene. 2016;**576**:701-707

[32] A. Poli, I. Finore, I. Romano, A. Gioiello, L. Lama, B. Nicolaus, Microbial Diversity in Extreme Marine Habitats and Their Biomolecules, Microorganisms, 5 (2017).

[33] C.D. Vavourakis, A.S. Andrei, M. Mehrshad, R. Ghai, D.Y. Sorokin, G. Muyzer, A metagenomics roadmap to the uncultured genome diversity in hypersaline soda lake sediments, 6 (2018) 168.

[34] Stetter KO. Hyperthermophilic procaryotes. FEMS microbiology reviews. 1996;**18**:149-158

[35] Xie W, Wang F, Guo L, Chen Z, Sievert SM, Meng J, et al. Comparative metagenomics of microbial communities inhabiting deep-sea hydrothermal vent chimneys with contrasting chemistries. The ISME journal. 2011;**5**:414-426

[36] Anantharaman K, Breier JA, Dick GJ. Metagenomic resolution

of microbial functions in deep-sea hydrothermal plumes across the Eastern Lau Spreading Center. The ISME journal. 2016;**10**:225-239

[37] Urich T, Lanzen A, Stokke R, Pedersen RB, Bayer C, Thorseth IH, et al. Microbial community structure and functioning in marine sediments associated with diffuse hydrothermal venting assessed by integrated meta-omics. Environmental microbiology. 2014;**16**:2699-2710

[38] Petersen JM, Zielinski FU, Pape T, Seifert R, Moraru C, Amann R, et al. Hydrogen is an energy source for hydrothermal vent symbioses. Nature. 2011;**476**:176-180

[39] Anantharaman K, Breier JA, Sheik CS, Dick GJ. Evidence for hydrogen oxidation and metabolic plasticity in widespread deep-sea sulfur-oxidizing bacteria. Proceedings of the National Academy of Sciences of the United States of America. 2013;**110**:330-335

[40] Brazelton WJ, Nelson B, Schrenk MO. Metagenomic evidence for h(2) oxidation and h(2) production by serpentinite-hosted subsurface microbial communities. Frontiers in microbiology. 2012;**2**:268

[41] Romano C, D'Imperio S, Woyke T, Mavromatis K, Lasken R, Shock EL, et al. Comparative genomic analysis of phylogenetically closely related Hydrogenobaculum sp. isolates from Yellowstone National Park. Applied and environmental microbiology. 2013;**79**:2932-2943

[42] Sorokin VA, Gelfand MS, Artamonova II. Evolutionary dynamics of clustered irregularly interspaced short palindromic repeat systems in the ocean metagenome. Applied and environmental microbiology. 2010;**76**:2136-2144

[43] Anderson RE, Brazelton WJ, Baross JA. Using CRISPRs as a metagenomic tool to identify microbial hosts of a diffuse flow hydrothermal vent viral assemblage. FEMS microbiology ecology. 2011;**77**:120-133

[44] Price PB. A habitat for psychrophiles in deep Antarctic ice. Proceedings of the National Academy of Sciences of the United States of America. 2000;**97**:1247-1251

[45] Cowan DA, Russell NJ, Mamais A, Sheppard DM. Antarctic Dry Valley mineral soils contain unexpectedly high levels of microbial biomass. Extremophiles: life under extreme conditions. 2002;**6**:431-436

[46] Simon C, Wiezer A, Strittmatter AW, Daniel R. Phylogenetic diversity and metabolic potential revealed in a glacier ice metagenome. Applied and environmental microbiology. 2009;**75**:7519-7526

[47] Mikucki JA, Priscu JC. Bacterial diversity associated with Blood Falls, a subglacial outflow from the Taylor Glacier, Antarctica. Applied and environmental microbiology. 2007;**73**:4029-4039

[48] Bond PL, Druschel GK, Banfield JF. Comparison of acid mine drainage microbial communities in physically and geochemically distinct ecosystems. Applied and environmental microbiology. 2000;**66**:4962-4971

[49] Baker BJ, Banfield JF. Microbial communities in acid mine drainage. FEMS microbiology ecology. 2003;**44**: 139-152

[50] Bruneel O, Duran R, Casiot C, Elbaz-Poulichet F, Personne JC. Diversity of microorganisms in Fe-As-rich acid mine drainage waters of Carnoules, France. Applied and environmental microbiology. 2006;**72**:551-556

[51] He Z, Xiao S, Xie X, Zhong H, Hu Y, Li Q, et al. Molecular diversity

of microbial community in acid mine drainages of Yunfu sulfide mine. Extremophiles: life under extreme conditions. 2007;**11**:305-314

[52] S. Xiao, X. Xie, J. Liu, Microbial communities in acid water environments of two mines, China, Environmental pollution (Barking, Essex: 1987), 157 (2009) 1045-1050.

[53] Dick GJ, Andersson AF, Baker BJ, Simmons SL, Thomas BC, Yelton AP, et al. Community-wide analysis of microbial genome sequence signatures. Genome biology. 2009;**10**:R85

[54] Huang LN, Kuang JL, Shu WS. Microbial Ecology and Evolution in the Acid Mine Drainage Model System. Trends in microbiology. 2016;**24**:581-593

[55] Bruneel O, Pascault N, Egal M, Bancon-Montigny C, Goni-Urriza MS, Elbaz-Poulichet F, et al. Archaeal diversity in a Fe-As rich acid mine drainage at Carnoules (France). Extremophiles: life under extreme conditions. 2008;**12**:563-571

[56] Golyshina OV, Timmis KN. Ferroplasma and relatives, recently discovered cell wall-lacking archaea making a living in extremely acid, heavy metal-rich environments. Environmental microbiology. 2005;**7**:1277-1288

[57] Kamekura M. Diversity of extremely halophilic bacteria. Extremophiles: life under extreme conditions. 1998;**2**:289-295

[58] N.M. Mesbah, J. Wiegel, Life at extreme limits: the anaerobic halophilic alkalithermophiles, Annals of the New York Academy of Sciences, 1125 (2008) 44-57.

[59] Mesbah NM, Wiegel J. Life under multiple extreme conditions: diversity and physiology of the halophilic alkalithermophiles. Applied and environmental microbiology. 2012;**78**:4074-4082

[60] Narasingarao P, Podell S, Ugalde JA, Brochier-Armanet C, Emerson JB, Brocks JJ, et al. De novo metagenomic assembly reveals abundant novel major lineage of Archaea in hypersaline microbial communities. The ISME journal. 2012;**6**:81-93

[61] Anton J, Rossello-Mora R, Rodriguez-Valera F, Amann R. Extremely halophilic bacteria in crystallizer ponds from solar salterns. Applied and environmental microbiology. 2000;**66**:3052-3057

[62] Plominsky AM, Delherbe N, Ugalde JA, Allen EE, Blanchet M, Ikeda P, et al. Metagenome sequencing of the microbial community of a solar saltern crystallizer pond at cahuil lagoon, chile. Genome announcements. 2014;**2**

[63] M. Kambourova, I. Tomova, I. Boyadzhieva, N. Radchenkova, E. Vasileva-Tonkova, Unusually High Archaeal Diversity in a Crystallizer Pond, Pomorie Salterns, Bulgaria, Revealed by Phylogenetic Analysis, Archaea (Vancouver, B.C.), 2016 (2016) 7459679.

[64] Fernandez AB, Ghai R, Martin-Cuadrado AB, Sanchez-Porro C, Rodriguez-Valera F, Ventosa A. Prokaryotic taxonomic and metabolic diversity of an intermediate salinity hypersaline habitat assessed by metagenomics. FEMS microbiology ecology. 2014;**88**:623-635

[65] Pikuta EV, Hoover RB, Tang J. Microbial extremophiles at the limits of life. Critical reviews in microbiology. 2007;**33**:183-209

[66] Dumorne K, Cordova DC, Astorga-Elo M, Renganathan P. Extremozymes: A Potential Source for Industrial Applications. Journal of microbiology and biotechnology. 2017;**27**:649-659

[67] Demirjian DC, Moris-Varas F, Cassidy CS. Enzymes from extremophiles. Current opinion in chemical biology. 2001;**5**:144-151

[68] Raddadi N, Cherif A, Daffonchio D, Neifar M, Fava F. Biotechnological applications of extremophiles, extremozymes and extremolytes. Applied microbiology and biotechnology. 2015;**99**:7907-7913

[69] Lentzen G, Schwarz T. Extremolytes: Natural compounds from extremophiles for versatile applications. Applied microbiology and biotechnology. 2006;**72**:623-634

[70] Arakawa T, Yamaguchi R, Tokunaga H, Tokunaga M. Unique Features of Halophilic Proteins. Current protein & peptide science. 2017;**18**:65-71

[71] Dalsgaard PW, Larsen TO, Christophersen C. Bioactive Cyclic Peptides from the Psychrotolerant Fungus Penicillium algidum. The Journal of Antibiotics. 2005;**58**:141

[72] Stierle AA, Stierle DB, Patacini B. The Berkeleyamides, Amides from the Acid Lake Fungus Penicillum rubrum. Journal of Natural Products. 2008;**71**:856-860

[73] Stierle AA, Stierle DB, Kelly K. Berkelic Acid, A Novel Spiroketal with Selective Anticancer Activity from an Acid Mine Waste Fungal Extremophile. The Journal of Organic Chemistry. 2006;**71**:5357-5360

[74] Jančič S, Frisvad JC, Kocev D, Gostinčar C, Džeroski S, Gunde-Cimerman N. Production of Secondary Metabolites in Extreme Environments: Food- and Airborne Wallemia spp. Produce Toxic Metabolites at Hypersaline Conditions. PLOS ONE. 2016;**11**:e0169116

[75] Mende DR, Aylward FO, Eppley JM, Nielsen TN, DeLong EF. Improved Environmental Genomes via Integration of Metagenomic and Single-Cell Assemblies. Frontiers in microbiology. 2016;**7**

[76] Gomariz M, Martínez-García M, Santos F, Rodriguez F, Capella-Gutiérrez S, Gabaldón T, et al. From community approaches to single-cell genomics: the discovery of ubiquitous hyperhalophilic Bacteroidetes generalists. The ISME journal. 2014;**9**:16

[77] Rashid M, Stingl U. Contemporary molecular tools in microbial ecology and their application to advancing biotechnology. Biotechnology advances. 2015;**33**:1755-1773

[78] Raes J, Foerstner KU, Bork P. Get the most out of your metagenome: computational analysis of environmental sequence data. Current opinion in microbiology. 2007;**10**:490-498

[79] Simon C, Daniel R. Metagenomic analyses: past and future trends. Applied and environmental microbiology. 2011;**77**:1153-1161

[80] Ferrer M, Beloqui A, Timmis KN, Golyshin PN. Metagenomics for mining new genetic resources of microbial communities. Journal of molecular microbiology and biotechnology. 2009;**16**:109-123

[81] Moore-Connors JM, Dunn KA, Bielawski JP, Van Limbergen J. Novel Strategies for Applied Metagenomics. Inflammatory bowel diseases. 2016;**22**:709-718

[82] Sudarikov K, Tyakht A, Alexeev D. Methods for The Metagenomic Data Visualization and Analysis. Current issues in molecular biology. 2017;**24**:37-58

[83] Mirete S, Morgante V, Gonzalez-Pastor JE. Functional metagenomics of extreme environments. Current opinion in biotechnology. 2016;**38**:143-149

[84] De Santi C, Altermark B, Pierechod MM, Ambrosino L, de Pascale D, Willassen NP. Characterization of a cold-active and salt tolerant esterase identified by functional screening of Arctic metagenomic libraries. BMC biochemistry. 2016;**17**:1

[85] Troeschel SC, Drepper T, Leggewie C, Streit WR, Jaeger KE. Novel tools for the functional expression of metagenomic DNA. Methods in molecular biology (Clifton, N.J.). 2010;**668**:117-139

[86] Liebl W, Angelov A, Juergensen J, Chow J, Loeschcke A, Drepper T, et al. Alternative hosts for functional (meta) genome analysis. Applied microbiology and biotechnology. 2014;**98**:8099-8109

[87] Park SY, Kim GJ. Screening of functional promoter from metagenomic DNA for practical use in expression systems. Methods in molecular biology (Clifton, N.J.). 2010;**668**:141-152

[88] Bitok JK, Lemetre C, Ternei MA, Brady SF. Identification of biosynthetic gene clusters from metagenomic libraries using PPTase complementation in a Streptomyces host. FEMS microbiology letters. 2017;**364**

[89] Schmidt EW, Donia MS. Chapter 23. Cyanobactin ribosomally synthesized peptides--a case of deep metagenome mining. Methods in enzymology. 2009;**458**:575-596

[90] Iqbal HA, Feng Z, Brady SF. Biocatalysts and small molecule products from metagenomic studies. Current opinion in chemical biology. 2012;**16**:109-116

[91] Nguyen T, Ishida K, Jenke-Kodama H, Dittmann E, Gurgui C, Hochmuth T, et al. Exploiting the mosaic structure of trans-acyltransferase polyketide synthases for natural product discovery and pathway dissection. Nature biotechnology. 2008;**26**:225-233

[92] Donia MS, Ravel J, Schmidt EW. A global assembly line for cyanobactins. Nature chemical biology. 2008;**4**:341-343

[93] Feng Z, Kallifidas D, Brady SF. Functional analysis of environmental DNA-derived type II polyketide synthases reveals structurally diverse secondary metabolites. Proceedings of the National Academy of Sciences of the United States of America. 2011;**108**:12629-12634

[94] Reynolds PE. Structure, biochemistry and mechanism of action of glycopeptide antibiotics, European journal of clinical microbiology & infectious diseases: official publication of the European Society of Clinical. Microbiology. 1989;**8**:943-950

[95] Donadio S, Sosio M. Biosynthesis of glycopeptides: prospects for improved antibacterials. Current topics in medicinal chemistry. 2008;**8**:654-666

[96] Banik JJ, Brady SF. Cloning and characterization of new glycopeptide gene clusters found in an environmental DNA megalibrary. Proceedings of the National Academy of Sciences of the United States of America. 2008;**105**:17273-17277

[97] Jaspars M. The origins of cyanobactin chemistry and biology. Chemical communications (Cambridge, England). 2014;**50**:10174-10176

[98] Leikoski N, Liu L, Jokela J, Wahlsten M, Gugger M, Calteau A, et al. Genome mining expands the chemical diversity of the cyanobactin family to include highly modified linear peptides. Chemistry & biology. 2013;**20**:1033-1043

[99] Martins J, Leao PN, Ramos V, Vasconcelos V. N-terminal protease gene phylogeny reveals the potential for novel cyanobactin diversity in cyanobacteria. Marine drugs. 2013;**11**:4902-4916

[100] Long PF, Dunlap WC, Battershill CN, Jaspars M. Shotgun cloning and heterologous expression of the patellamide gene cluster as a strategy to achieving sustained metabolite production, Chembiochem: a European. journal of chemical biology. 2005;**6**:1760-1765

[101] Donia MS, Ruffner DE, Cao S, Schmidt EW. Accessing the hidden majority of marine natural products through metagenomics, Chembiochem: a European. journal of chemical biology. 2011;**12**:1230-1236

[102] Hertweck C, Luzhetskyy A, Rebets Y, Bechthold A. Type II polyketide synthases: gaining a deeper insight into enzymatic teamwork. Natural product reports. 2007;**24**:162-190

[103] Wawrik B, Kerkhof L, Zylstra GJ, Kukor JJ. Identification of unique type II polyketide synthase genes in soil. Applied and environmental microbiology. 2005;**71**:2232-2238

[104] Piel J. Biosynthesis of polyketides by trans-AT polyketide synthases. Natural product reports. 2010;**27**:996-1047

[105] Zimmermann K, Engeser M, Blunt JW, Munro MH, Piel J. Pederin-type pathways of uncultivated bacterial symbionts: analysis of o-methyltransferases and generation of a biosynthetic hybrid. Journal of the American Chemical Society. 2009;**131**:2780-2781

[106] Stepanauskas R. Single cell genomics: an individual look at microbes. Current opinion in microbiology. 2012;**15**:613-620

[107] T. Woyke, D.F.R. Doud, The trajectory of microbial single-cell sequencing, 14 (2017) 1045-1054.

[108] Blainey PC. The future is now: single-cell genomics of bacteria and archaea. FEMS microbiology reviews. 2013;**37**:407-427

[109] Rinke C, Lee J, Nath N, Goudeau D, Thompson B, Poulton N, et al. Obtaining genomes from uncultivated environmental microorganisms using FACS-based single-cell genomics. Nature protocols. 2014;**9**:1038-1048

[110] Xu Y, Zhao F. Single-cell metagenomics: challenges and applications. Protein & cell. 2018;**9**:501-510

[111] Nurk S, Bankevich A, Antipov D, Gurevich AA, Korobeynikov A, Lapidus A, et al. Assembling single-cell genomes and mini-metagenomes from chimeric MDA products. Journal of computational biology: a journal of computational molecular cell biology. 2013;**20**:714-737

[112] Kodzius R, Gojobori T. Single-cell technologies in environmental omics. Nature methods. 2016;**576**:701-707

[113] Yilmaz S, Singh AK. Single cell genome sequencing. Current opinion in biotechnology. 2012;**23**:437-443

[114] Hedlund BP, Dodsworth JA, Murugapiran SK, Rinke C, Woyke T. Impact of single-cell genomics and metagenomics on the emerging view of extremophile "microbial dark matter". Extremophiles: life under extreme conditions. 2014;**18**:865-875

[115] Baker BJ, Tyson GW, Webb RI, Flanagan J, Hugenholtz P, Allen EE, et al. Lineages of acidophilic archaea revealed by community genomic analysis. Science (New York, N.Y.). 2006;**314**:1933-1935

[116] Baker BJ, Comolli LR, Dick GJ, Hauser LJ, Hyatt D, Dill BD, et al. Enigmatic, ultrasmall, uncultivated Archaea. Proceedings of the National Academy of Sciences of the United States of America. 2010;**107**:8806-8811

Chapter 5

Halocins, Bacteriocin-Like Antimicrobials Produced by the Archaeal Domain: Occurrence and Phylogenetic Diversity in *Halobacteriales*

Afef Najjari, Hiba Mejri, Marwa Jabbari, Haitham Sghaier, Ameur Cherif and Hadda-Imene Ouzari

Abstract

Members of extremely halophilic archaea, currently consisting of more than 56 genera and 216 species, are known to produce their specific bacteriocin-like peptides and proteins called halocins, synthesized by the ribosomal pathway. Halocins are diverse in size, consisting of proteins as large as 35 kDa and peptide "microhalocins" as small as 3.6 kDa. Today, about fifteen halocins have been described and only three genes, halC8, halS8 and halH4, coding C8, S8 and H4 halocins respectively have been identified. In this study, a total of 1858 of complete and nearly complete genome sequences of Halobacteria class members were retrieved from the IMG and Genbank databases and then screened for halocin encoding gene content, based on the BLASTP algorithm. A total of 61 amino acid sequences belonging to three halocins classes (C8, HalH4 and S8) were identified within 15 genera with the abundance of C8 class. Phylogenetic analysis based on amino acids sequences showed a clear segregation of the three halocins classes. Halocin S8 was phylogenetically more close to HalH4. No clear segregation on species and genera levels was observed based on halocin C8 analysiscontrary to HalH4 based analysis. Collectively, these results give an overview on halocins diversity within halophilic archaea which can open new research topics that will shed light on halocins as marker for haloarchaeal phylogentic delineation.

Keywords: archaea, bioinformatics, diversity, halocins, phylogeny

1. Introduction

Microorganisms of the third domain of life, Archaea, have been cultivated and described for more than 100 years [1], however, they have been first assigned to the Bacteria domain because of their great phenotypic similarities. In the late 1970s, Carl Woese and his collaborators, recognized the Archaea as the third domain of life on earth based on molecular phylogenetic analyses [2]. The dichotomous (eukaryotic/prokaryotic) classification was no longer valid, leading to a reclassification of organisms as three separate domains: Eucarya (*Eukaryotes*), Archaea and Bacteria (bacteria) [3]. Archaea share several similarities with the other two domains of life. They are similar to size-level bacteria, organization of their chromosomes, absence of nucleus and organelles, presence of polycistronic transcription units and use of Shine-Dalgarno sequences for the initiation of the translation. In addition, it was shown that their metabolic proteins are essentially bacterial in nature following analysis of many complete genomes of Archaea [2]. Archaea also share similarities with the Eucarya domain, such as the proteins involved in key informational processes such as replication [4], transcription, translation [5, 6], DNA repair [7, 8], mRNA degradation and proteolysis. Translation in Archaea has eukaryotic initiation and elongation factors, and their transcription involves TATA binding protein and TFIIB [9].

The biotopes colonized by these microorganisms, are supposed to approach to the primitive terrestrial atmosphere (high salinity or pH, devoid of O_2, rich in H_2 and CO_2 constituting the raw materials for the production of methane) [10]. They present spectacular adaptations, especially in extreme environments. We distinguish: (i) Thermophilic Archaea: living at high temperatures (60–80°C) (ii) Hyperthermophilic Archaea: living at very high temperatures (up to 121°C); (iii) Psychrophilic Archaea: prefering low temperatures (below 15°C) [11]; (iv) Halophilic Archaea: colonizing very saline environments (3–5 M NaCl) such as the Dead Sea [12, 13]; (v) Acidophilic Archaea: living at low pH (as low as pH 1 and dying at pH 7) and Alkaliphilic Archaea: thriving at high pH (up to 9) [14].

2. Taxonomy of the archaeal domain

The first phylogenetic study based on the comparison of the 16S rDNA gene sequences coding for the small subunit, separated the first founding members of Archaea into two taxa, one grouping methanogenic species and those living under conditions of extreme salinity, the other containing species living at very high temperatures and at acidic pH [15]. Ten years later, analyses on a larger taxonomic group led to the division of the Archaea kingdom into two groups: (i) *Crenarchaeota*, which is composed exclusively of microorganisms living at very high temperatures, and (ii) *Euryarchaeota*, a heterogeneous group of species with different phenotypes (methanogenic species, species living at very high temperatures, moderate temperatures or at high salt concentrations) [3]. Fifteen years later and thanks to metagenomic analyses, two phyla, *Thaumarchaeota* and *Korarchaeota* were established based on the results of genomic comparison of two uncultivable strains, *Candidatus cenarchaeum symbosium* and *Candidatus korarchaeum cryptofilum*, with genomic traits belonging to both phyla *Crenarchaeota* and *Euryarchaeota* [16]. On the other hand, the symbiont *Nanoarchaeum equitans*, occupying cells of the host *Ignicoccus hospitalis*, showed even more genomic divergence with the other members of the *Crenarchaeota* and was therefore the first member of the phyla *Nanoarchaeota*.

Today, we count more of 15 phyla in the reign of Archaea, some of them having been grouped in superphylum. One distinguishes the superphylum TACK, proposed in 2011 and of which the eukaryotes would have evolved according to the theory of the eocyte, grouping *Thaumarchaeota*, *Aigarchaeota*, *Crenarchaeota* and *Korarchaeota* phyla [17]. This superphylum has been joined by recently proposed phyla: *Bathyarchaeota*, *Geoarchaeota* and *Lokiarchaeota* [18]. Another superphylum, DPANN, was proposed in 2013 and includes *Diapherotrites*, *Parvarchaeota*, *Aenigmarchaeota*, *Nanoarchaeota* and *Nanohaloarchaeota* phyla (**Figure 1**) [19]. Finally, the phyla *Woesearchaeota* and *Pacearchaeota*, described in 2016, were grouped in the DPANN superphylum (**Figure 1**).

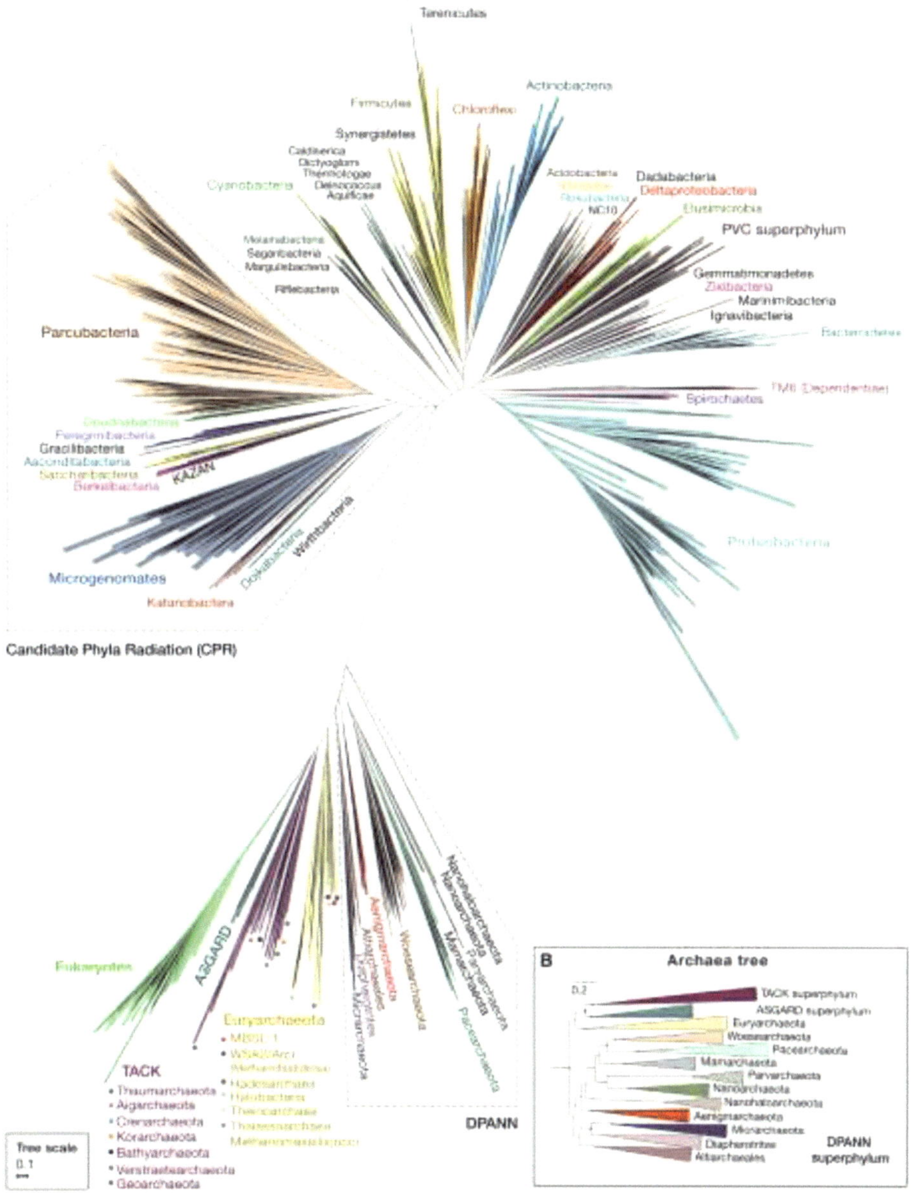

Figure 1.
Representation of the tree of life based on SSU rDNA gene sequences of the three domain of life [23].

3. Antimicrobial potential of extremely halophilic archaea

Halophilic archaea were the first members of archaea found to produce bacteriocins-like proteins known as halocins. The first studies date from the beginning of 1980s with experiments demonstrating the presence of antagonistic interactions between halophilic archaeal strains isolated from the Alicante salt in Spain [20]. Today, about fifteen halocins have been described and only three genes, *halC8*, *halS8* and *halH4*, coding C8, S8 and H4 halocins, respectively, have been identified [21, 22]. Yet, no three-dimensional structural data of archaeocins are available in public databases.

3.1 Halocins

Halocins, bacteriocins-like peptides and proteins produced by extremely halophilic archaea, were first discovered in 1982 by F. Rodriguez Valera [24, 25]. They are classified according to their size into two major classes: high molecular mass (protein, > 10 kDa) and low molecular mass (peptide, ≤10 kDa) called microhalocins [26, 27]. It has been shown that halocins are effective against *Haloarchaea* and *Crenarchaea* such as *Sulfolobus* spp. and *Methanosarcina thermopila*, and thus act across the main subdivision of the archaeal domain. These compounds represent a general class of antiarchaeal toxins and there is no confirmation about the inhibition of bacteria [26, 27]. Production of halocin is a practically universal feature of archaeal halophilic rods [27]. Although several halocins were identified, only some of them have been characterized and purified.

3.1.1 Microhalocins

These halocins are composed of a peptide with size below or in the range of 10 kDa. Seven halocins have been characterized including HalS8, HalR1, HalC8, HalU1, HalH6, Sech7a and Sech10. They are hydrophobic and retain their activity in the absence of salt and can be stored at 4°C. They are relatively insensitive to heat and organic solvents [28].

3.1.1.1 Halocin S8 (HalS8)

HalS8 is the first characterized microhalocin with 36 amino acids (3580 Da), it is synthesized by the uncharacterized S8a haloarchaea [29]. Halocin S8 showed a narrow inhibitory spectrum and can only inhibit *Halobacterium salinarum* NRC817, *Halobacterium* GRB and *Haloferax gibbonsii* [29]. It can be desalted and it is heat resistant. Its activity is resistant to trypsin but sensitive to proteinase K and is undetectable in the transition to the stationary phase [29, 30]. The *halS8 gene* is *encoded* on a ~200-kbp megaplasmid [29].

3.1.1.2 Halocin HalR1 (HalR1)

Halocin R1, the second characterized microhalocin, is produced by *Halobacterium salinarum* GN101, a strain isolated from solar salt marsh in Mexico [31]. Initially HalR1 was described with a molecular weight of 6.2 kDa [32] and later on, it was shown that the HalR1 peptide is composed of 38 amino acids [24, 29]. Like HalS8, the activity is not affected by desalting and is resistant to acids, bases, organic solvents DNase and RNase, and against some proteases such as papain, trypsin or thermolysin, but it is sensitive to proteinase K, pronase P and elastase [20, 32].

3.1.1.3 Halocin C8 (HalC8)

Halocin C8 is produced by *Natrinema* sp. AS7092, a strain isolated from the large Chaidan Salt Lake, China (7.44 kDa, 76 amino acids) [28]. It is a unique polypeptide with an isoelectric point of 4.4 [33]. Its activity is retained after desalting, boiling and frozing [33]. Halocin C8 has a very broad spectrum of activity against several species and genera of *Halobacteriales* members including *Natronobacterium gregoryi*, *Nbt*. comb. nov and *Natronomonas pharaonis* [28]. The *halC8* gene encodes both halocin C8 and its immunity protein HalI.

3.1.1.4 Halocin A4 (HalU1)

Halocin A4, also called also halocin U1, is produced by an uncharacterized haloarchaea strain isolated from a Tunisian saltern [34]. Its molecular weight is 7.435 Da, as determined by the spectrometric mass, and is both acidic (pH = 4.14) and hydrophobic (eluent at ~85% acetonitrile) [26]. Halocin A4 has been reported to inhibit the growth of crenarchaeal *Sulfolobus* sp. strains [26]. G*ene* encoding HalA4 is *located* on a 300 kpb megaplasmid, pHM300 (NC_017943) [29].

3.1.1.5 Halocin H6 (HalH6)

Halocin H6 is produced by *Haloferax gibbonsii* Ma 2.39 species [27]. Its activity is resistant to trypsin. Stabilities of this peptide were studied and have shown that HalH6 can be desalted and it retained its activity after heat treatment up to 10 min at 100°C [27]. Halocin H6 is considered as a bactericidal substance which causes cell lysis and the specific target of HalH6 is the Na^+/H^+ antiport [27, 35].

3.1.1.6 Halocin Sech7a

Halocin Sech7a was excreted by the extremely halophilic haloarchaeon Sech7a, isolated from brine samples of Secovlje solar salterns crystallizers in Slovenia [36]. Sech7a is about 11 kDa. It is stable over a wide pH range and is heat labile at temperatures above 80°C. Its optimal activity was observed in the early exponential phase growth at 45°C. It loses activity under low salt conditions, but its activity can be restored after dialysis against initial saline conditions [36].

3.1.1.7 Halocin SH10

Halocin SH10 is produced by *Natrinema* sp. BTSH10, a strain isolated from the Kanyakumari salt marsh, Tamil Nadu, India [37]. The optimal production of halocin SH10 is at 42°C, pH 8.0 and 3 M NaCl at the stationary phase. In this context, it was reported that the activity is lost under acidic conditions [37]. Production of SH10 is influenced by the carbon source composition of the medium — *Natrinema* sp. BTSH10 could produce maximal halocin in the presence of beef [37].

3.1.2 Protein halocins

This class comprises halocins composed of proteins greater than 10 kDa in size. Currently, there are two characterized protein halocins, HalH1 and HalH4, in the range of 30 to 35 kDa [28].

3.1.2.1 Halocin H4

Halocin H4, produced by *Haloferax mediterranei* R4 (ATCC 33500) was isolated from a Spanish solar salt pond in Alicante. It is the first halocin that was studied [20]. The optimal activity was detected at the midpoint between exponential and stationary phases [20]. Halocin H4 is sensitive to proteases, high temperature and desalting. HalH4 has an antimicrobial activity against other haloarchaeons. It Interacts with the membrane of the target cells where it causes permeability changes that result in an ionic imbalance leading to cell lysis and death [21, 38]. The *halH4* gene, encoding halocin activity, is located on the pHM300 megaplasmid, a single polypeptide of 34.9 kDa.

3.1.2.2 Halocin H1

Halocin H1 is produced by *Haloferax mediterranei* M2a (previously known as *H. mediterranei* Xia3) isolated from salt ponds in Santa Pola (Alicante, Spain) [20]. Halocin H1 is a single 31 kDa polypeptide characterized by a broad inhibitory spectrum among *Halobacteriales* members. HalH1 activity is temperature and salt dependant. It is stable at 50°C only and requires a salt concentration of 1.5 M to maintain its activity [38, 39]. Optimum activity was observed at mid-exponential phase. The sensitivity to proteases and the gene encoding activity were not determined yet.

3.2 Applications of halocins

Some studies reported the role of halocins in a variety of environmental, industrial and biotechnological applications ***(REFERENCES?). However, this topic is poorly documented and somewhat controversial. One of these applications is the use of halocin producing strains in the textile industry during the tanning process characterized by high salinity concentration, halocins could inhibit the growth of pathogenic microbes affecting the quality of products. [7, 10]. Moreover, some halocins have also been reported for biomedical and therapeutic uses, for example, Halocin H7 has been shown to inhibit the Na+/H+ antiport in *Haloarchaea*, can be used as a treatment to reduce the injuries caused when ischemic organ transplantation is re-infused [35]. The therapeutic potential of halocins needs more research on their physical structures and their modes of action. On the other hand, halocins are known also to have a potential application in food industry as preservative agents by controlling the growth of haloarchaea in salted food products [40].

4. Materials and methods

Here, we evaluated the evolutionary relationship between bacteriocin- like-producing haloarchaea members based on comparisons of their amino acid sequences retrieved from annotated genomes sequences deposited in the IMG database [41].

4.1 Database search of halocin gene clusters

Schematic workflow of the methodology employed of amino acid sequences retrieving and phylogenetic assessment is illustrated in **Figure 2**. The methodology consisted of: first, complete and nearly complete genome sequences of *Halobacteriales* members were retrieved from IMG database. Then, in *silico* screening for gene sequences encoding halocins was done based on the BLASTP algorithm with default parameters [42]. All redundant and low-quality sequences were eliminated from datasets.

4.2 Phylogenetic reconstruction

Multiple sequences alignment of retrieved amino acid sequences were performed using ClustalW [43]. The evolutionary history was inferred using the Unweighted pair group method with arithmetic mean (UPGMA) method [44] implemented in MEGA X [45, 46]. The optimal tree with the sum of branch length = 18.99 is shown. Percentages of replicate trees in which the associated taxa clustered together in the bootstrap test (1000 replicates) are shown next to the branches. The tree is drawn to scale, with branch lengths in the same units as those of the evolutionary distances used to infer the phylogenetic tree. The evolutionary distances were computed using the Poisson correction method [47] and are in the

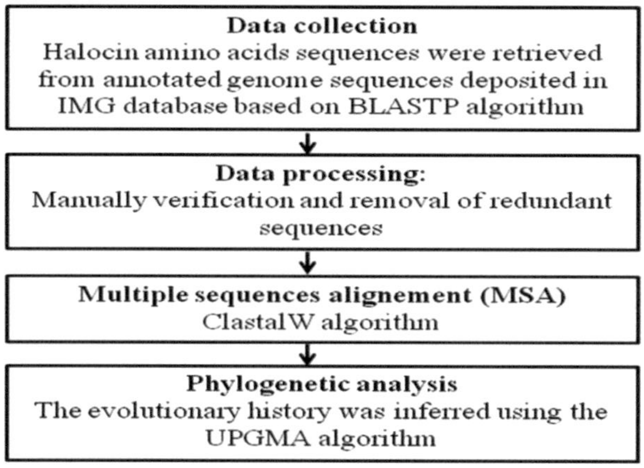

Figure 2.
Schematic workflow of the methodology employed for amino acids sequences retrieving and phylogenetic assessment.

units of the number of amino acid substitutions per site. All ambiguous positions were removed for each sequence pair (pairwise deletion option). In the final dataset, a total of 405 positions was obtained.

5. Results

5.1 Amino acid sequence of halocins

A total of 1858 of complete and nearly complete genome sequences of *Halobacteriales class* members were retrieved from the IMG database and creened for halocin encoding gene based on the BLASTP algorithm with default parameters [42].

A total of 61 amino acid sequences were retrieved from 15 genera belonging to *Halobaceria* class including *Natrinema*, *Haloferax*, *Haloterrigena*, *Natronorubrum*, *Halobacterium*, *Haladaptatus, Halorubrum, Halococcus, Halopiger*, *Natrialba, Halolamina*, *Natronococcus, Haloarcula, Halapricum and Halorussus*. Furthermore, some other unclassified halohilic archaea were present as well, including uncultured halophilic archaeon, halophilic archaeon sp. DL31 and Haloarchaeon S8 (**Table 1**).

Results showed that some species present more than one copy for halocin encoding genes. In fact, three (n = 3) classes of halocins were identified in this study (**Table 1**).

The first class is halocin C8-like bacteriocin domain (HalC8), the best known bacteriocin like sequences in archaea, it has been demonstrated to be produced from a ProC8 precursor, targeted to the membrane by the Tat pathway, and cleaved by an unknown mechanism to yield the active mature peptide HalC8 and an immunity protein HalI, protecting the producing strain against its own AMP [22]. HalC8 was identified in all species except *Natrialba aegyptia* DSM 13077. Indeed, *Natrinema* genus members appears to be more represented in terms of C8-like bacteriocin production. It's worth noting that among the six officially described species within the *Natrinema* genus, five are described in the current analysis. Several studies report the production of HalC8 and/or the presence of the *halC8* gene among *Natrinema* species isolated from different geographical origins like Chaidan Salt Lake in Qinghai province, China [33], Ichekaben salterns Chotts and sebkhas in algeria [48, 49].

The second class is halocin H4 (HalH4) identified in *Haloferax mediterranei* strain ATCC 33500, *Natrialba aegyptia and Natrinema gari* JCM 14663 species.

Taxonomy	Genus	Species level	Class of Halocins
Domain: *Archaea* Kingdom: *Euryarchaeota* Phylum: *Euryarchaeota* Class: *Halobacteria* Order: *Halobacteriales* Family: *Halobacteriaceae*	*Natrinema*	*Natrinema pellirubrum* 157 JCM 10476	Halocin C8-like bacteriocin
		Natrinema sp. J7-2	
		Natrinema pallidum DSM 3751	
		Natrinema pellirubrum DSM 15624	
		Natrinema altunense 1A4-DGR	
		Natrinema altunense JCM 12890	
		Natrinema sp. J7-1	
		Natrinema altunense AJ2	
		Natrinema atunense 4.1	
		Natrinema gari JCM 14663	Halocin C8-like bacteriocin and Halocin H4
	Haloferax	*Haloferax mediterranei* ATCC 33500	Halocin H4
		Haloferax mediterranei R-4	
		Haloferax lucentense DSM 14919	Halocin C8-like bacteriocin
		Haloferax sp. ATCC BAA-646	
		Haloferax volcanii DS2	
		Haloferax alexandrinus JCM 10717	
		Haloferax sp. ATCC BAA-645	
		Haloferax larsenii CDM 5	
		Haloferax larsenii JCM 13917	
	Haloterrigena	*Haloterrigena thermotolerans*	Halocin H4
		Haloterrigena salifodinae ZY19	Halocin C8-like bacteriocin
		Haloterrigena jeotgali A29	
		Haloterrigena mahii H13	
		Haloterrigena salina JCM 13891	
		Haloterrigena sp. P1A	
		Haloterrigena turkmenica WANU15	
	Natronorubrum	*Natronorubrum tibetense* DSM 13204	Halocin C8-like bacteriocin
		Natronorubrum tibetense GA33	
		Natronorubrum sediminis CGMCC 1.8981	

Taxonomy	Genus	Species level	Class of Halocins
Domain: *Archaea* Kingdom: *Euryarchaeota* Phylum: *Euryarchaeota* Class: *Halobacteria* Order: *Halobacteriales* Family: *Halobacteriaceae*	*Halobacterium*	*Halobacterium* sp. DL1	Halocin C8-like bacteriocin
		Halobacterium salinarum DSM 670	
		Halobacterium salinarum DSM 671	
		Halobacterium salinarum DSM 6692	
		Halobacterium salinarum DSM 3754	
		Halobacterium salinarum DSM 668	
	Haladaptatus	*Haladaptatus paucihalophilus* DX253	
		Haladaptatus sp. R4	
		Haladaptatus paucihalophilus DSM 18195	
	Halorubrum	*Halorubrum lacusprofundi* R1S1	
		Halorubrum trapanicum CBA1232	
	Halococcus	*Halococcus* sp. 197A	
		Halococcus salifodinae DSM 8989	
	Halopiger	*Halopiger* sp. IIH3	
	Natrialba	*Natrialba aegyptia* DSM 13077	
	Halolamina	*Halolamina pelagica* CGMCC	
	Natronococcus	*Natronococcus occultus* SP4	
	Haloarcula	*Haloarcula salaria* H5-DGR	
	Halapricum	*Halapricum salinum* CBA1105	
	Halorussus	*Halorussus amylolyticus YC93*	
	Halophilic archaeon	halophilic archaeon sp. DL31	
	uncultured halophilic archaeon	uncultured halophilic archaon J07HX5	
		uncultured haloarchaeon J07ABHX67	
		Uncultured Halobacteriaceae archaea SG1_71_5	
	Haloarchaeon S8a	*Haloarchaeon S8a*	Halocin S8

Table 1.
Classes of Halocins identified by in silico analysis of all genomes of halophilic archaea domain available in IMG database.

HalH4 was first characterized from *H. mediterranei* isolated from solar saltern lakes of Spain [50]. HalH4 is a 40 kDa protein with an N-terminal 46 aa leader peptide which is cleaved off leaving a 313 aa mature halocin [51].

The third class is halocin S8, a microhalocin of 36 amino acids (3580 Da) initially purified from an unidentified haloarchaeal strain S8a, isolated from the Great Salt Lake (Utah, 109 United States) [52].

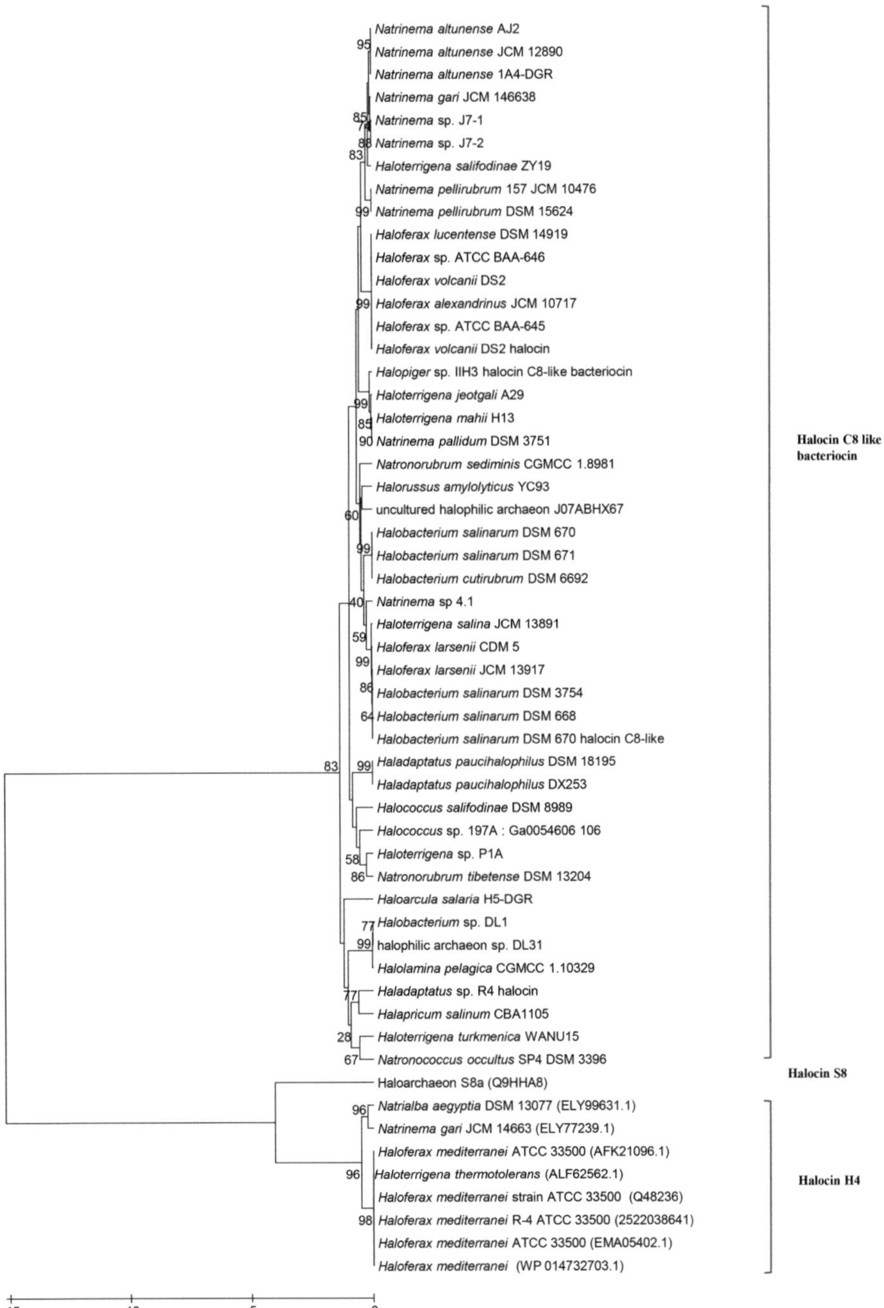

Figure 3.
Phylogenetic tree of halocin amino acid sequences of halophilic archaeal species. The evolutionary history was inferred using the UPGMA algorithm implemented in MEGA X software. Numbers at the nodes indicate the percentage of occurrence in 100 bootstrapped trees (bootstrap values > 50% are shown).

5.2 Phylogenetic analysis

Phylogenetic analysis of retrieved halocin peptide sequences was conducted and the result is illustrated in **Figure 3**. Results showed a clear segregation of the three halocins classes (C8, H4 and S8), where halocin S8 is phylogenetically more close to HalH4. Furthemore, no clear separation of species was observed based on HalC8 amino acids sequences analyses. HalC8 was detected in 12 genera belonging to three orders of *Halobacteria* class [53]: *Natrialbales* (*Natrinema*, *Haloterrigena*, *Natronorubrum*, *Halopiger*); *Haloferacales* (*Haloferax*, *Halolamina*) and *Halobacteriales* (*Halorussus*, *Halobacterium*, *Haladaptatus*, *Halococcus*, *Haloarcula*, *Halapricum*) and uncultured halophilic archaeon (J07ABHX67) phylogenetically related to species *Halorussus amylolyticus* YC93. The halocin S8 was detected only in the strain Haloarchaeon S8a (Q9HHA8). HalH4 is identified in *Natrialbales* (*Natrinema gari* JCM 14663 (ELY77239.1), *Haloterrigena thermotolerans*, *Natrialba aegyptia* DSM 13077) and *Haloferacales* (*Haloferax*).

It's worth noting that HalH4/HalC8 halocins were identified in *Haloferax, Haloterrigena* and *Natrinema* genera with only the species *Natrinema gari* JCM 14663 (ELY77239.1) being able to produce the two classes in the same time. Earlier studies reported that several described halocins, with broad inhibitory properties, are derived from *Haloferax* and *Natrinema* strains [24, 28, 45, 51, 54] and it has been suggested that halocin production may explain their dominance in some saline ecosystems [54, 55].

6. Conclusion

On the basis of our *in silico* analyses, we can conclude that halocin production is considered as a general feature of some members of halophilic archaea, particularly members of *Natrialbales* and *Haloferacales* orders with the occurrence of Halocin C8-like production. This group can thrive in saline ecosystems in which several other microorganisms are not able to live. Thus, the dominance of certain species isolated in some saline ecosystems could be attributed to halocin production as a mechanism of competition between microrganisms. This chapter will open new research lines that will shed light on halocins as marker for haloarchaeal phylogentic delineation.

Conflict of interest

We have *no conflict of interest* to declare.

Author details

Afef Najjari[1*], Hiba Mejri[1], Marwa Jabbari[2,3], Haitham Sghaier[2,4], Ameur Cherif[4] and Hadda-Imene Ouzari[1]

1 Faculté des Sciences de Tunis, Université de Tunis El Manar, LR03ES03 Microorganismes et Biomolécules Actives, 2092, Tunis, Tunisia

2 Laboratory "Energy and Matter for Development of Nuclear Sciences" (LR16CNSTN02), National Center for Nuclear Sciences and Technology (CNSTN), Sidi Thabet Technopark, 2020, Tunisia

3 Biochemistry and Molecular Biology, Code UR13ES34 Research Unit, Faculty of Science of Bizerte, University of Carthage, 7021 Zarzouna, Tunisia

4 ISBST, BVBGR-LR11ES31, Biotechpole of Sidi Thabet, Univ. Manouba, 2020 Ariana, Tunisia

*Address all correspondence to: najjariafef@gmail.com; afef.najjari@fst.utm.tn

References

[1] Cavicchioli R. *archaea* — timeline of the third domain. (2011). Nature Reviews Microbiology. 9(1):51-61.

[2] Woese CR, Fox GE. (1977). The concept of cellular evolution. J Mol Evol. 10(1):1-6.

[3] Woese CR, Kandler O, Wheelis ML (1990). Towards a natural system of organisms: proposal for the domains Archaea, Bacteria, and Eucarya. Proc Natl Acad Sci U S A. 87(12):4576-9.

[4] Barry ER, Bell SD. DNA Replication in the Archaea (2006). Microbiol Mol Biol Rev. 70(4):876-87.

[5] Londei P (2005). Evolution of translational initiation: new insights from the archaea. FEMS Microbiol Rev. 29(2):185-200.

[6] Steitz TA (2008). A structural understanding of the dynamic ribosome machine. Nature Reviews Molecular Cell Biology. 9(3):242-53.

[7] Fujikane R, Ishino S, Ishino Y, Forterre P (2010). Genetic analysis of DNA repair in the hyperthermophilic archaeon, Thermococcus kodakaraensis. Genes Genet Syst. 85(4):243-57.

[8] Kelman Z, White MF (2005). Archaeal DNA replication and repair. Curr Opin Microbiol. 8(6):669-76.

[9] Soppa J (2006). From genomes to function: haloarchaea as model organisms. Microbiology. 152(Pt 3):585-90.

[10] Karr J (2006). Seven foundations of biological monitoring and assessment. Biologia Ambientale. 20:7-18.

[11] Van de Vossenberg JL, Driessen AJ, Konings WN (1998). The essence of being extremophilic: the role of the unique archaeal membrane lipids. Extremophiles. 2(3):163-70.

[12] Rothschild LJ, Mancinelli RL (2001). Life in extreme environments. Nature. 409(6823):1092-101.

[13] Margesin R, Schinner F (2001). Potential of halotolerant and halophilic microorganisms for biotechnology. Extremophiles.5(2):73-83.

[14] Quatrini R, Johnson DB (2016). Acidophiles: Life in Extremely Acidic Environments. Caister Academic Press; 2016. 300 p.

[15] Fox GE, Stackebrandt E, Hespell RB, Gibson J, Maniloff J, Dyer TA, et al (1980). The phylogeny of prokaryotes. Science. 209(4455):457-63.

[16] Brochier-Armanet C, Boussau B, Gribaldo S, Forterre P (2008). Mesophilic *Crenarchaeota*: proposal for a third archaeal phylum, the *Thaumarchaeota*. Nat Rev Microbiol. 6(3):245-52.

[17] Guy L, Ettema TJG (2011). The archaeal 'TACK' superphylum and the origin of eukaryotes. Trends in Microbiology. 19(12):580-7.

[18] Castelle CJ, Wrighton KC, Thomas BC, Hug LA, Brown CT, Wilkins MJ, et al. (2015). Genomic Expansion of Domain Archaea Highlights Roles for Organisms from New Phyla in Anaerobic Carbon Cycling. Current Biology. 25(6):690-701.

[19] MacLeod G, Bozek DA, Rajakulendran N, Monteiro V, Ahmadi M, Steinhart Z, et al (2019). Genome-Wide CRISPR-Cas9 Screens Expose Genetic Vulnerabilities and Mechanisms of Temozolomide Sensitivity in Glioblastoma Stem Cells. Cell Rep. 27(3):971-986.

[20] Rodriguez-Valera F, Juez G, Kushner D (2011). Halocins: salt-dependent bacteriocins produced by extremely

halophilic rods. Canadian Journal of Microbiology. 28:151-4.

[21] Cheung J, Danna KJ, O'Connor EM, Price LB, Shand RF (1997). Isolation, sequence, and expression of the gene encoding halocin H4, a bacteriocin from the halophilic archaeon Haloferax mediterranei R4. J Bacteriol. 179(2):548-51.

[22] Sun C, Li Y, Mei S, Lu Q, Zhou L, Xiang H (2005). A single gene directs both production and immunity of halocin C8 in a haloarchaeal strain AS7092. Mol Microbiol. 57(2):537-49.

[23] Castelle CJ, Banfield JF. Major New Microbial Groups Expand Diversity and Alter our Understanding of the Tree of Life (2018). Cell. 172(6):1181-97.

[24] O'Connor E, Shand R (2002). Halocins and sulfolobicins: The emerging story of archaeal protein and peptide antibiotics. Journal of industrial microbiology & biotechnology. 28:23-31.

[25] Li Y, Xiang H, Tan H (2002). Halocin: protein antibiotics produced by extremely halophilic archaea.

[26] Haseltine C, Hill T, Montalvo-Rodriguez R, Kemper SK, Shand RF, Blum P (2001). Secreted euryarchaeal microhalocins kill hyperthermophilic crenarchaea. J Bacteriol. 183(1):287-91.

[27] Torreblanca M, Meseguer I, Rodriguez-Valera F (1989). Halocin H6, a Bacteriocin from Haloferax gibbonsii. Microbiology-sgm. 135:2655-61.

[28] Shand R, Leyva K (2008). Archaeal Antimicrobials: An Undiscovered Country. Archaea: New Models for Prokaryotic Biology. Caister Academic Press.

[29] Price LB, Shand RF (2000). Halocin S8: a 36-Amino-Acid Microhalocin from the Haloarchaeal Strain S8a. J Bacteriol. 182(17):4951-8.

[30] Shand R, Price LB, O'Connor E (1999). Halocins: Protein antibiotics from hypersaline environments. In: Oren A (Ed), Microbiology and Biogeochemistry of Hypersaline Environments.CRC Press, Boca Raton, FL. 295-306.

[31] Ebert K, Goebel W, Rdest U, Surek B (1986). Genes and genome structures in the archaebacteria. Syst Appl Microbiol. 7, 30-35. 1986.

[32] Rdest U, Sturm M (1987). Bacteriocins from halobacteria. In: Burgess R, editor. Protein purification: micro to macro. New York, N.Y: Alan R. Liss, Inc.; pp. 271-278.

[33] Li Y, Xiang H, Liu J, Zhou M, Tan H (2003). Purification and biological characterization of halocin C8, a novel peptide antibiotic from Halobacterium strain AS7092. Extremophiles. 7(5):401-7.

[34] Shand R (2006). 29 Detection, Quantification and Purification of Halocins: Peptide Antibiotics from Haloarchaeal Extremophiles. Methods in Microbiology. 35:703-18.

[35] Meseguer I, Torreblanca M, Konishi T (1995). Specific inhibition of the halobacterial Na+/H+ antiporter by halocin H6. J Biol Chem. 270(12):6450-5.

[36] Pasić L, Velikonja BH, Ulrih NP (2008). Optimization of the culture conditions for the production of a bacteriocin from halophilic archaeon Sech7a. Prep Biochem Biotechnol. 38(3):229-45.

[37] Karthikeyan P, Bhat SG, Chandrasekaran M (2013). Halocin SH10 production by an extreme haloarchaeon Natrinema sp. BTSH10

isolated from salt pans of South India. Saudi Journal of Biological Sciences. 20(2):205-12.

[38] Perez AM (2000). Growth Physiology of Haloferax Mediterranei R4 and Purification of Halocin H4. Northern Arizona University; 156 p.

[39] Meseguer I, Rodriguez-Valera F (1986). Effect of Halocin H4 on Cells of Halobacterium halobium. Microbiology-sgm. 132:3061-8.

[40] Charlesworth J, Burns BP (2016). Extremophilic adaptations and biotechnological applications in diverse environments. AIMS Microbiology. 2(3):251.

[41] Markowitz EM, Goldberg LR, Ashton MC, Lee K (2012). Profiling the "Pro-Environmental Individual": A Personality Perspective. Journal of Personality. 80(1):81-111.

[42] Altschul SF, Gish W, Miller W, Myers EW, Lipman DJ (1990). Basic local alignment search tool. Journal of Molecular Biology. 215(3):403-10.

[43] Thompson JD, Higgins DG, Gibson TJ (1994). CLUSTAL W: improving the sensitivity of progressive multiple sequence alignment through sequence weighting, position-specific gap penalties and weight matrix choice. Nucleic Acids Res. 22(22):4673-80.

[44] Sokal RR, Michener CD (1958). A statistical method for evaluating systematic relationships. 28, 1409-1438.

[45] Kumar S, Stecher G, Li M, Knyaz C, Tamura K (2018). MEGA X: Molecular Evolutionary Genetics Analysis across Computing Platforms. Mol Biol Evol. 35(6):1547-9.

[46] Sneath PHA, Sokal RR (1973). Numerical Taxonomy: The Principles and Practice of Numerical Classification. San Francisco: W.H.Freeman & Co Ltd; 1973. 588 p.

[47] Zuckerkandl E, Pauling L (1965). Molecules as documents of evolutionary history. J Theor Biol. 8(2):357-66.

[48] Imadalou-Idres N, Carré-Mlouka A, Vandervennet M, Yahiaoui H, Jean P, Rebuffat S (2013). Diversity and Antimicrobial Activity of Cultivable Halophilic Archaea from Three Algerian Sites. 7:1057-69.

[49] Quadri I, Hassani I, L'haridon S, Chalopin M, Hacene H, Jebbar M (2016). Characterization and antimicrobial potential of extremely halophilic archaea isolated from hypersaline environments of the Algerian Sahara. Microbiological Research. 186.

[50] Quesada E, Ventosa A, Rodriguez-Valera F, Ramos-Cormenzana A (1982). Types and properties of some bacteria isolated from hypersaline soils. Journal of Applied Bacteriology. 53(2):155-61.

[51] Besse A, Jean P, Rebuffat S, Carré-Mlouka A (2015). Antimicrobial peptides and proteins in the face of extremes: Lessons from archaeocins. Biochimie. 118.

[52] Rinke C, Schwientek P, Sczyrba A, Ivanova NN, Anderson IJ, Cheng J-F, et al (2013). Insights into the phylogeny and coding potential of microbial dark matter. Nature.499(7459):431-7.

[53] Gupta RS, Naushad S, Baker S (2015). Phylogenomic analyses and molecular signatures for the class Halobacteria and its two major clades: a proposal for division of the class *Halobacteria* into an emended order *Halobacteriales* and two new orders, *Haloferacales* ord. nov. and Natrialbales ord. nov., containing the novel families *Haloferacaceae* fam. nov. and Natrialbaceae fam. nov. Int J Syst Evol Microbiol. 65(Pt 3):1050-69.

[54] Atanasova N, Pietilä M, Oksanen H (2013). Diverse antimicrobial interactions of halophilic archaea and bacteria extend over geographical distances and cross the domain barrier. MicrobiologyOpen. 2.

[55] Kis-Papo T, Oren A (2000). Halocins: are they involved in the competition between halobacteria in saltern ponds? Extremophiles. 4(1):35-41.

Chapter 6

Tree Species and Precipitation Effect on the Soil Microbial Community Structure and Enzyme Activities in a Tropical Dry Forest Reserve

Michelle Jovanne Rivera-Rivera, Veronica Acosta-Martínez and Elvira Cuevas

Abstract

We examined the soil microbial community structure and soil enzyme potential within three dominant tree species at a tropical dry forest during five months. Changes within microbial community in response to sampling periods and tree species were evaluated using fatty acid methyl-ester and enzymes potential. We found that both tree species and precipitation determined microbial community structure and enzyme potential. This is the first study that provides insight into the soil microbial community at Guánica Dry Forest, a valuable contribution that will help elucidate strategies for better management and protection of the soil biota of the area.

Keywords: dry soil, soil enzyme activity, microbial diversity, EL-FAME, phosphatases

1. Introduction

Forest ecosystems are a fundamental part of the biosphere as they govern global primary production and biogeochemical cycling [1]. The dynamics of decomposition and nutrient cycling is driven by soil microbial communities [2] and their enzymatic potential [3]. Assessments of soil microbial community composition are often coupled with the assessment of soil processes (C storage and nutrient cycling) through different enzymatic activities to better address the ecosystem function [4]. Some of the principal enzyme activities assayed in soils are involved in carbon cycling (β-glucosidase, α-galactosidase) with potential importance for C sequestration [5]. Other enzymes such as β-glucosaminidase are important in the decomposition of more complex compounds such as chitin [6, 7] Phosphorus cycling enzymes (acid and alkaline phosphatases and phosphodiesterase) are important since most soils are P deficient [8] and microbial enzyme activity (EA) plays a vital role in the availability of P in soils.

Both microbial communities and enzyme activities are sensitive to biotic and abiotic disturbances. For example, low soil moisture may result in low enzyme activity [9] and rainfall pulses may result in pulses of microbial growth which may lead to pulsed secretion of soil enzymes promoting a temporary increase of soil enzyme activity [10].

Understanding the dynamics of forest soil ecosystem depends on elucidating the contribution of individual plant species to the soil biota and the process that they regulate [11]. Determining and quantifying plant species effects under natural conditions can be difficult due to environmental noise and the interaction among species present in an area [12]. Little is known about the effects of dominant vegetation on the soil biogeochemical processes such as enzyme activities [13]. Even less is known about the plant-soil-microbial interactions that take place in extreme environments such as tropical dry forest.

Nearly 42% of tropical forests around the world are seasonally dry plant communities, where around half of the Central American and Caribbean land area is characterized by a tropical or subtropical dry forest climate. In the Coastal Plateau of the Guánica Dry Forest (GDF), an UNESCO/MAB Biosphere Reserve, dwarfed trees grow isolated from one another in the cracks of the calcareous platform, forming individual islands of fertility [14]. A lack of interspecific competition is observed, as tolerance to environmental stress and scarcity of space for establishment make it difficult for aboveground and belowground competition, also contributing to the evenness of tree species found in this area [14]. The substrate is derived from limestone made from marine deposits that vary throughout the forest from deep alluvial fans to exposed fractured limestone with shallow soil pockets [15]. This naturally occurring plant community provides the ideal conditions to determine how specific tree species affect the soil microbial community composition and enzymatic potential in a dry forest.

In order to understand how trees impact soil microbial communities, a five-month study was conducted at the GDF. We selected three dwarfed, isolated tree species (a pantropical species and two native species) that are highly distributed among the forest, hypothesizing that these trees may harbour different microbial community structure and activities. The tree species selected complied with the following requirements: (1) that trees were growing in cracks isolated from other trees by exposed rock and (2) that their litter and belowground substrate originated from their own residue decomposition [16]. Additionally, this forest experiences bimodal and pulsed precipitation patterns [17, 18] that may contribute in the alteration of the microbial dynamics and nutrient turnover of the forest. Our objectives were: (1) to determine if tree species traits had an effect on the soil microbial community structure and activities and (2) to determine the effects of sampling period on the soil microbial community structure and activities.

2. Methods

2.1 Study site

The study was located in the coastal association of the Guánica Dry Forest latitude 66°53′30″W longitude 17°58′0″N [16–18]. In this area of the forest, dwarf individual adult trees from different species are separated by exposed rocks which prevent overlap among trees forming "monospecific tree islands" (**Figure 1A** and **B**). These "monospecific tree islands" have little or no mineral soil (**Figure 1A**), with a highly organic substrate composed of shallow monospecific litter and humus (**Figure 1C** and **D**), which varies according to variations in the ground relief and season from 2 to 8 cm depth [16]. These characteristics make this forest an ideal system to study tree species effects as there are no confounding effects of overlapping roots from other species. We chose three trees from three species, previously tagged and studied, that grow from 100 m to approximately 300 m from the coast.

Figure 1.
Detailed description of study site. (A) Tree species that has established its growth in the cracks and crevices of the calcareous rock. Black arrow demonstrates a new soil pocket that is being formed as a by-product of the decomposition of the tree's litter. (B and C) Tabebuia heterophylla *tree that has established on the calcareous rock at the Guánica Dry Forest; here, we can observe small litter pockets that have formed in the cracks of the rock. (D) Figure demonstrates that the substrate collected is humus with a vast quantity of fragmented litter, and dry adventitious roots.*

The three tree species selected are: *Tabebuia heterophylla* (*DC.*) Briton, Ann, a facultative deciduous species; *Pisonia albida* (Heimerl) Briton ex Scandal, an obligate deciduous species and *Ficus citrifolia* Mill., Gard., a facultative deciduous species. The trees are distributed more or less randomly across this area and are dominant species of the coastal region [17], no spatial distribution has been detected. During this study, the highest precipitation was reported during the month of August 2011 (212 mm) and the lowest precipitation was reported for the month of October 2011 (85.09 mm). The total accumulated rainfall was 738 mm.

2.2 Selected soil substrate analyses

Total organic matter, soil texture, pH and moisture analyses were completed. Soil total organic matter was determined by the ignition method. Soil texture was determined by the pipette method [19]. Soil pH was determined in a 1:5 (soil:water) mixture using an Orion 420A pH Meter [20]. Soil moisture was determined by weighing and drying 1 g of soil at 105°C during 24 h [20].

2.3 EL-FAME analysis

Microbial community structure was evaluated by the ester-linked (EL)-FAME method as described by [21]. A total of 3 g of field moist soil was used for each sample, and four steps were completed: (a) saponification and methylation of ester-linked fatty acids were performed by incubating the 3 g of soil in 15 ml of 0.2 M KOH in methanol at 37°C during 1 h—samples were vortexed every 10 min during the incubation period and, after the incubation period was completed, 3 ml of 1.0 acetic acid was added to neutralize the mixture's pH; (b) partitioning of the FAMEs into an organic phase was achieved by adding 10 ml of hexane and centrifuging the preparations at 480 ×*g* for 10 min; (c) the hexane layer was transferred to a clean glass tube and evaporated under a N^2 stream and (d) FAMEs were suspended in 0.5 ml of 1:1 hexane:methyl-tert butyl ether and transferred to GC vials for analysis. Extractions were analysed as described by [4]. A 6890 GC Series II (Hewlett Packard, Wilmington, DE, USA) equipped with a flame ionization detector and fused silica capillary column (25 m × 0.2 mm) with ultra-high purity H^2 as the gas carrier was used to analyse the extractions. The temperature program was ramped from 170 to 250°C at 5°C min^{-1} as described. The fatty acids were identified and quantified by comparing the retention times and peak areas to MIDI standards. The MIDI software provides FAME relative peak areas (percentage) based on the total FAMEs in a sample (based on the Aerobe method of the MIDI system). FAME concentrations (nmol g^{-1} soil) were calculated by comparing peak areas to an analytical standard (19:0, Sigma Chemical Co., St. Louis, MO) calibration curve. The FAMEs are described by the number of C atoms, followed by a colon, the number of double bonds and then by the position of the first double bond from the methyl (ω) end of the molecule. Cis isomers are indicated by c and branched fatty acids are indicated by the prefixes *i* and *a* for iso and anteiso, respectively. Other notations are Me for methyl, OH for hydroxy and cy for cyclopropane.

2.4 Soil enzyme activity

Study of enzyme activities was performed as described in [4, 6]. The activities of enzymes relevant in C cycling (β-glucosidase), C and N cycling (β-glucosaminidase), P cycling (alkaline phosphatase, acid phosphatase, phosphodiesterase) and in the S cycle (arylsulphatase) were assayed using 0.5 g of air-dried soil (<2 mm). Duplicate replicates and one control were used for all the soils that were tested; furthermore, the appropriate substrate was used for each assay and reactions were incubated for 1 h at 37°C at their optimal pH as described in [4]. For the controls, the substrate was added after the 1-h incubation period and subtracted from a sample control value. Enzyme activity is expressed in mg p-nitrophenol (PN) released in kg^{-1} soil h^{-1}.

2.5 Statistical analysis

All data analysed with JMP software were checked for normality and transformed to log10. To determine differences in soil FAMEs, enzyme activity and selected soil properties due to tree species and sampling periods, canonical discriminate analysis (CDA) was conducted with the JMP program. The first and second canonical discriminate functions were utilized to determine the distribution of enzyme activity and FAMEs as influenced by each tree species and by sampling period. SigmaPlot 10 software was used to conduct two-way analysis of variance (2-way ANOVA). Two-way analysis of variance was used to establish the effect of plant species and sampling period on soil community structure, enzyme activities and selected soil properties.

3. Results

3.1 Selected soil and plant properties

Soil samples under all tree species were predominantly organic (65–85%). The mineral part of the soil is silty loam to silty clay loam (20–36% clay, 0.4–0.7% sand,

63–80% silt) depending on the tree location in the study area (data not shown). Soil pH was neutral to alkaline for all tree species (**Table 1**). Two-way ANOVA did not demonstrate an interaction between tree species and sampling months (**Table 2**). Available N varied significantly with regard to monthly sampling and not to tree species (**Table 2**). Available Ca and P varied significantly with regard to tree species (**Table 2**). Total N and C:N ratio varied significantly with regard to tree species.

3.2 Soil enzyme activities

Six enzyme activities (EAs) were assessed representing C (β-glucosidase), N (β-glucosaminidase), P (phosphodiesterase, alkaline and acid phosphatase) and S (arylsulphatase) cycling. There was a clear and significant separation in the soil enzyme activity between the tree species in this forest, while there was no distinct trend due to the sampling time (**Figure 2**). Most of the separation on the metabolic capacity of the soil according to these six EAs was observed along axis 1, which explained 87.22% of the variability. *Tabebuia heterophylla* aligned along CA 1 separating this species from *Pisonia albida* and *Ficus citrifolia* which aligned along CA 2 (eigenvalue 12.3). The activities of β-glucosidase, alkaline and acid phosphatase were more closely associated with *Tabebuia heterophylla* than with the other tree species while β-glucosaminidase activity was more closely associated with *Ficus citrifolia* trees.

Two-way analysis of variance was used to establish the effect of plant species and sampling period on soil enzyme activities. The only enzymes that were significantly affected by plant species were alkaline phosphatase ($F = 8.18$; $P < 0.001$), β-glucosidase ($F = 8.86$; $P < 0.001$) and β-glucosaminidase ($F = 5.97$; $P = 0.007$) (**Table 2**). Alkaline phosphatase was the only enzyme affected by monthly variations ($F = 4.375$; $P = 0.007$). Two-way ANOVA does not demonstrate significant interactions between tree species and monthly variations.

3.3 Microbial community structure (EL-FAME)

A total of 56 FAMEs were identified in the samples analysed and 13 FAMEs were used as indicators of different microbial groups as affected by tree species traits and sampling time (**Table 2**). According to CDA, significant differences ($P < 0.0001$) were detected in the FAME profiles of the soil microbial community structure as affected by tree species traits (**Figure 3A**). Canonical axis 1 explained 83% of the variation. The soil microbial communities under *Tabebuia heterophylla* were distinct from the microbial communities under *Ficus citrifolia* and *Pisonia albida*. FAMEs that contributed to the separation observed under *Tabebuia heterophylla* are the

	Tree Species						
ected Soil Properties	*F. citrifolia*		*P. albida*		*T. heterophyla*		p-value (α 0.05)
cid Phosphatase[a]	867.79 (221.68)	B	1472.65 (1229.14)	A	1237.61 (245)	AB	0.03
-glucosaminidase[a]	169.70 (42.28)	AB	296.90 (251.60)	A	116.78 (46)	B	0.00
Aryl sulphatase[a]	434.49 (231.98)	A	420.93 (216.88)	A	623.96 (498)	A	0.22
caline phosphatase[a]	1550.92 (689.80)	AB	1148.74 (576.70)	B	1950.45 (488)	A	0.02
hosphodiesterase[a]	721.32 (326.18)	A	404.28 (197.62)	B	774.60 (238)	A	0.00
B-glucosidase[a]	288.49 (60.40)	AB	255.95 (92.73)	B	560.45 (352)	A	0.00
Soil humidity (%)	72 (29.98)	A	112.57 (69.16)	A	88.18 (29.2)	A	0.16
Biomass	2606.84 (410.54)	A	2306.95 (678.03)	A	2746.88 (1061)	A	0.28
pH	8.02 (0.23)	A	7.69 (0.71)	A	7.79 (0.51)	A	0.20

Table 1.
Mean values of selected soil properties under three different tree species.

Variables	Species		Month		Species x Month	
	F	P	F	P	F	P
Soil Properties						
Soil Organic Matter	1.70	0.20	0.52	0.72	0.17	0.99
pH	10.31	<0.001	1.28	0.30	0.25	0.98
Soil Humidity	1.28	0.29	1.11	0.37	0.88	0.55
Total Available N	1.74	0.19	3.02	0.03	0.58	0.79
Total N	4.81	0.02	0.46	0.77	0.11	1.00
Total C	2.29	0.12	0.37	0.83	0.13	1.00
C/N	9.36	<0.001	0.87	0.50	0.45	0.88
Total Available Ca	6.16	0.01	0.95	0.45	0.79	0.61
Total Available P	6.59	0.00	0.59	0.68	0.36	0.93
Soil Enzymes						
Acid Phosphatase	2.63	0.09	1.12	0.37	1.04	0.43
B_glucosaminidase	5.97	0.01	0.90	0.48	1.26	0.30
Aryl Sulphatase	1.50	0.24	0.99	0.43	0.49	0.85
Alkaline Phosphatase	8.82	<0.001	4.38	0.01	0.77	0.64
Phosphodiesterase	11.27	<0.001	1.69	0.18	2.04	0.08
B-Glucosidase	8.86	<0.001	0.83	0.52	0.87	0.55
EL-FAMES						
iso15:0	21.60	<0.001	7.44	<0.001	1.16	0.36
antesio15:0	0.69	0.51	3.42	0.02	0.67	0.72
iso17:0	10.04	<0.001	2.49	0.06	0.81	0.60
antesio17:0	7.75	0.00	1.50	0.23	0.96	0.49
17:0 cy	8.57	0.00	0.74	0.57	0.21	0.99
19:0 cy	6.87	0.00	0.61	0.66	1.08	0.41
10Me16:0	9.35	<0.001	3.17	0.03	0.37	0.93
10Me17:0	9.65	<0.001	0.32	0.86	0.65	0.73
10Me18:0	2.29	0.12	0.85	0.51	0.85	0.57
18:1ω9c	9.95	<0.001	1.04	0.40	0.86	0.56
16:1ω5c	5.32	0.01	4.32	0.01	0.53	0.83
18:2ω6c	6.31	0.01	5.71	0.00	0.94	0.50
18:3 ω6c	8.50	0.00	7.88	<0.001	1.92	0.09
20:4ω6c	0.54	0.59	0.96	0.45	0.43	0.89
Total Bacteria	2.74	0.08	1.14	0.36	0.91	0.52
Total Fungi	12.82	<0.001	1.52	0.22	1.38	0.24
Fungi:Bacteria	8.85	<0.001	0.97	0.44	1.10	0.39

Table 2.
*Two-way analysis of variance for the effects of species (*Ficus citrifolia, Pisonia albida *and* Tabebuia heterophylla*) and sampling periods (July 2011, August 2011, September 2011, October 2011 and November 2011) on the soil properties, soil enzymes and EL-FAMEs in the humus soil substrate layer at the Guánica Dry Forest.*

Gram-positive (G+) markers *a*15:0 and *a*17:0, actinomycete 10Me18:0, protozoan marker 20:4ω6c and the fungal marker 18:3ω6c.

Ficus citrifolia and *Pisonia albida* grouped closer together but were still significantly different from each other as species, 95% confidence ellipse are clearly separated. FAMEs that contribute to differences in *Ficus citrifolia* vs. *Pisonia albida* are cy19:0 (Gram-negative, G−), 10Me16:0 or 10Me17:0 (actinomycete) and 18:2 ω6c (fungal marker). Differences in *Pisonia albida we*re due to *i*17:0 (G+), cy17:0 (G−), and 16:1 ω5c or 18:1ω9c (fungal markers).

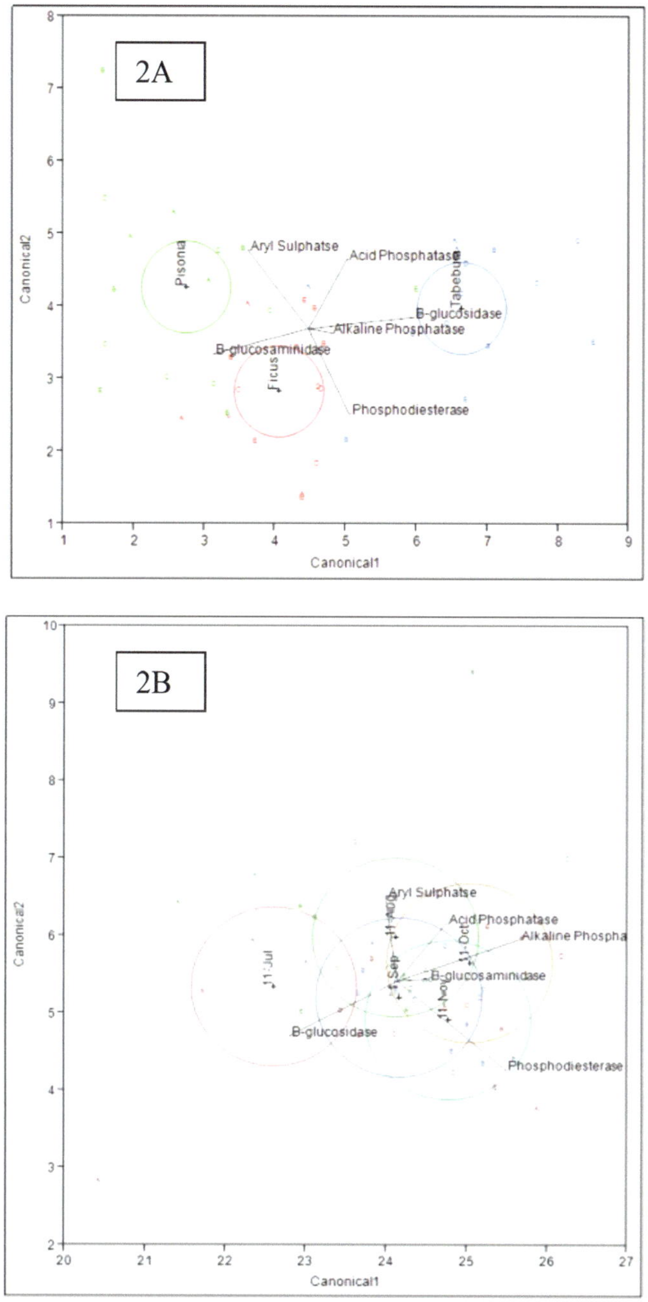

Figure 2.
*Canonical discriminate analysis of the soil enzyme activity at the Guánica Dry Forest. (A) canonical discriminate analysis of the soil enzyme activity as affected by tree species (Ficus citrifolia, Pisonia albida and Tabebuia heterophylla). Letters (A, B, C, D and E) represent sampling periods (July 2011, August 2011, September 2011, October 2011 and November 2011), respectively. The colours of the letters (pink, green and blue) represent tree species (*Ficus citrifolia, Pisonia albida *and* Tabebuia heterophylla*), respectively. The multivariate mean for each tree species is a coloured and labelled circle. The size of the circles corresponds to a 95% confidence limit for the mean. (B) canonical discriminate analysis of the soil enzyme activity as affected by sampling period during July to November 2011 at the Guánica Dry Forest of Puerto Rico. Letters (A, B, C, D and E) represent sampling periods (July 2011, August 2011, September 2011, October 2011 and November 2011), respectively. The colours of the letters (pink, green and blue) represent tree species (Ficus citrifolia, Pisonia albida and Tabebuia heterophylla), respectively. The multivariate mean for each month sampled is a coloured and labelled circle. The size of the circles corresponds to a 95% confidence limit for the mean.*

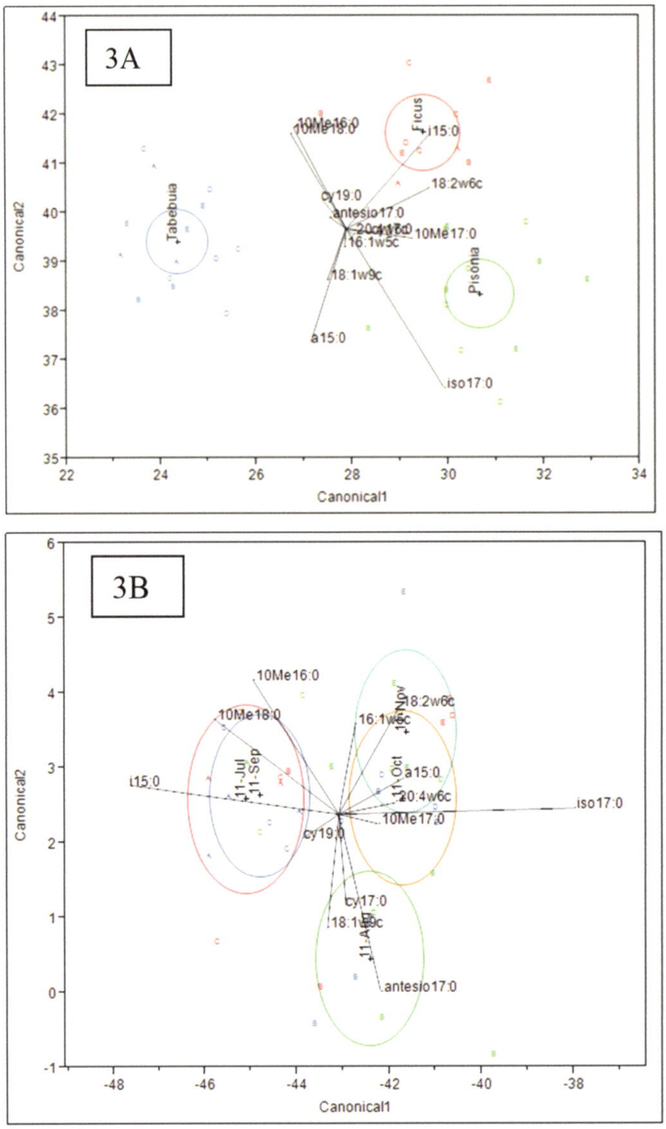

Figure 3.
*Canonical discriminate analysis of the soil FAMEs at the Guánica Dry Forest. (A) canonical discriminate analysis of the soil FAME profiles as affected by tree species (Ficus citrifolia, Pisonia albida and Tabebuia heterophylla). Letters (A, B, C, D and E) represent sampling periods (July 2011, August 2011, September 2011, October 2011 and November 2011), respectively. The colours of the letters (pink, green and blue) represent tree species (*Ficus citrifolia, Pisonia albida *and* Tabebuia heterophylla*), respectively. The multivariate mean for each tree species is a coloured and labelled circle. The size of the circles corresponds to a 95% confidence limit for the mean. (3B) canonical discriminate analysis of the soil FAME profiles as affected by sampling period during July to November 2011 at the Guánica Dry Forest of Puerto Rico. Letters (A, B, C, D and E) represent sampling periods (July 2011, August 2011, September 2011, October 2011 and November 2011), respectively. The colours of the letters (pink, green and blue) represent tree species (Ficus citrifolia, Pisonia albida and Tabebuia heterophylla), respectively. The multivariate mean for each month sampled is a coloured and labelled circle. The size of the circles corresponds to a 95% confidence limit for the mean.*

In addition to the effects of tree species on FAME profiles of the microbial community structure, there were also significant monthly variations ($p = 0.001$) according to CDA along CA1, which explained 68.81% of the variation (**Figure 3B**). The CDA revealed that July, September and August clustered together along axis 1. August was separated from other months due to the predominance of *a*17:0 (G+) and cy17:0 (G– marker). This cluster of these 3 months was characterized by the presence of the fungal marker 18:1ω9c, the G− marker cy19:0 and the actinomycete indicator 10Me16:0. The second cluster formed was composed of October and November due to the predominance of two G+ markers (*a*15:0 and *i*17:0), two actinomycete markers (10Me17:0 and 10Me18:0), a protozoan marker (20:4ω6c) and all the fungal markers (16:1ω5c, 18:3 ω6c and18:2ω6c).

Two-way analysis of variance was used to establish the effect of plant species and sampling period on soil community structure. The Gram-positive EL-FAMEs that were significantly affected due to tree species were iso15:0, iso17:0, antesio17:0, 10Me16:0 and 10Me17:0. Gram-negative markers affected by tree species were 17:0 cy and 19:0 cy and fungal markers were 18:1ω9c, 16:1ω5c, 18:2 ω6c and 18:3 ω6c. The Gram-positive EL-FAMEs that were significantly affected due to monthly variations were iso15:0, antesio15:0 and 10Me16:0. None of the Gram-negative markers were affected by monthly variations. Fungal markers 16:1ω5c, 18:2 ω6c and 18:3 ω6c were significantly affected by monthly variations. Two-way ANOVA did not demonstrate significant interactions between tree species and monthly variations (**Table 2**).

4. Discussion

4.1 Soil microbial communities in a dry forest as affected by tree species

The evaluated tree species had the ability of modifying soil microbial communities. We found that the relative abundance of 8 out of 13 markers was higher under *Ficus citrifolia* than the rest of the tree species. *Ficus citrifolia* is a facultative deciduous tree, which means that it mostly exchanges leaves and has massive litter fall during very dry periods. Higher numbers of FAME markers under this species may be indicative of idiosyncratic effects that may be stabilizing the microbial community.

Species traits such as leaf nutrients, leaf toughness and specific leaf area (SLA, cm^2/g) are known to contribute to the rate of decomposition of organic matter, which may drive the microbial community under each tree species [22–24]. Previous studies have determined that *Tabebuia heterophylla* leaves are tougher (350 N) and have lower specific leaf area (SLA) (90 cm^2/g) than those of *F. citrifolia* (175 N; 110 cm^2/g) and *Pisonia albida* (80 N; 100 cm^2/g) [16] making them more difficult to undergo decomposition. Soils under *Tabebuia heterophylla* presented a higher relative abundance of saprophytic fungal marker 18:3ω6c which may be indicative of lower rates of litter fragmentation and decomposition.

Differences in the relative abundance of fungal markers at the Guánica Dry Forest could also be due to the microclimate of each tree species. The amplitude of temperature fluctuations encountered in this forest varies among plant species, a factor that has been documented to affect arthropods in this system [16]. The canopies of both *Pisonia albida* and *Ficus citrifolia* generally are taller than the tree canopy of *Tabebuia heterophylla*. This difference in height contributes to differences in temperature, water throughfall and soil moisture. Idiosyncratic tree species characteristics have the ability of modifying the amplitude of daily temperature at the Guánica Dry Forest [16] and these modifications will also affect the fungal community structure that is present under each tree species.

4.2 Soil microbial communities in a dry forest as affected by monthly variations

As found in our study, microbial community structure was influenced by monthly wet/dry events (**Figure 3B**). Previous studies have described significant responses of the soil microbial communities to wet/dry events [25]. Our results point towards differential responses between sporadic and continuous rainfall events. The fungal markers showed lower abundance during July and higher abundance during October and November. Saprophytic fungal marker (18:3 ω6c, 18:2 ω6c) and the mycorrhiza marker 16:1 ω5c were more susceptible to monthly rainfall variations ($p < 0.0001$) than any other microbial group. As a consequence of rainfall pulses, adventitious roots were observed in the soil substrate, which serves as a surface area for the establishment of arbuscular mycorrhizal fungi. Although arbuscular mycorrhizal fungi enhance P absorption in root systems, we did not find any correlation between available P and mycorrhizal marker 16:1 ω5c.

Although fungal markers showed greater monthly fluctuations, the G+ bacterial marker *i*15:0 also responded significantly to monthly variations, specifically to water input. Dijkman [26] postulated that FAME marker (*i*15:0) is found in sulphate-reducing bacteria, which could explain our result. For example, the accumulation of water due to high rain pulses could make the soil habitat an anaerobic substrate, contributing to the proliferation of anaerobic and sulphate-reducing bacteria. Sulphate-reducing bacteria are widely spread in anaerobic habitats and play crucial roles in S and C mineralization [27]. For instance, during November, samples were collected after 2 days of continuous rain and the abundance of FAME marker *i*15:0 increased responding to rainfall pulses, which could support the possibility that certain sulphate-reducing bacteria were represented by this FAME marker and this would be in agreement with the C and S cycling enzyme activity trends during the rainfall pulse.

4.3 Soil enzymatic activity

Our results show that enzyme activity is highly dependent on the soil microbial community structure of each tree species. Tree species was a strong modulator of soil enzyme activity when compared to monthly climatic variations. Although all enzymes tested were active under each tree species, activity of certain enzymes was higher under specific tree species. For instance, acid phosphatase, alkaline phosphatase and β-glucosidase activity was higher under *Tabebuia heterophylla,* and β-glucosaminidase activity was the highest reported for *Pisonia albida.*

Tree species idiosyncratic traits affect not only the structure of microbial communities but also their enzyme activities. Alkaline phosphatase under *Pisonia albida* correlated with many microbial groups (Gram-positive, Gram-negative, actinobacteria, protozoa and saprophytic and arbuscular fungal markers). Alkaline phosphatase under *Ficus citrifolia* only correlated with Gram-positive, saprophytic and arbuscular fungi and finally alkaline phosphatase under *T. heterophylla* correlated with Gram-negative, actinobacteria and saprophytic fungi. Each enzyme activity correlated with specific microbial FAMEs, which varied depending on the tree species. Our results agree with [28] where they found that phosphatase enzymes correlated with higher numbers of fatty acids [28].

Microbial phosphatase activity is crucial for the supply of inorganic phosphate in this system. Soils at the Guánica Dry Forest are P limited due to the high amount of carbonates from the underlying calcareous substrate. In this study, phosphatases were the most active enzymes, microbial communities are allocating resources to balance the P deficiency of the Guánica soils. When P deficiency is present in a system, increased phosphatase activity occurs as a response to P starvation [29]. A mechanism that compensates soil P deficiency is the inoculation of the arbuscular mycorrhizae, which are known to enhance plant P availability via the production of phosphatase enzymes [30]. We identified similar concentrations of the arbuscular mycorrhizae marker 16:1 ω5c for *Tabebuia heterophylla* and *Ficus citrifolia.* Idiosyncratic effect was observed for each tree species; for example, *Ficus citrifolia* enhanced the activity of phosphodiesterase and *Tabebuia heterophylla* enhanced the activity of both acid and alkaline phosphatase. Wu et al. [31] found that *Poncirus trifoliate* seedlings that were inoculated with arbuscular mycorrhizae exhibited higher total and acid phosphatase activity and higher plant P content when compared to uninoculated seedlings. It is important to mention that other than arbuscular mycorrhizae, bacteria also produce phosphatase enzymes. We propose that microbial phosphatase activity is crucial for the availability of P at the Guánica Dry Forest Reserve.

5. Conclusions

Idiosyncratic effects of tree species coupled with extreme changes in water input contribute in shaping the soil microbial community structure and enzymatic activity at the Guánica Dry Forest. In this system, saprophytic fungi, arbuscular mycorrhizae and anaerobic Gram-positive sulphur-reducing bacteria seem to be more sensitive to rainfall pulses when compared to Gram-positive (including actinobacteria), Gram-negative and protozoan communities. Even though monthly variations play a significant role in microbial community structure, soil enzyme activities did not vary during the months sampled. Our findings demonstrate that, although mesoclimate is a determinant driver of ecosystems, tree species is a stronger modulator of the soil microbial dynamics at the Guánica Dry Forest. To our understanding, this is the first study that provides insight into the soil microbial community of the Guánica Dry Forest, a valuable contribution that will help elucidate strategies for better management and protection of the soil biota of the area.

Acknowledgements

The project was funded by NSF Grant HRD 0734826 and is a contribution of the Centre of Applied Tropical Ecology and Conservation of the University of Puerto Rico in collaboration with the USDA-ARS in Lubbock, Texas. We appreciate the support of Mr. Larry Diaz, laboratory coordinator, and students from the Ecosystems Processes and Function laboratory of the University of Puerto Rico, Río Piedras Campus, that participated in the field collection of soil samples; Mr. Jon Cotton, laboratory technician, and students from Dr. Veronica Acosta-Martinez laboratory at the USDA-ARS Lubbock Texas, that participated in sample processing, and Mr. Dean Holder from the USDA-ARS Lubbock Texas, who offered training in the determination of soil texture. Special thanks to Dr. Francisco Calderon from the USDA-ARS Akron Colorado for kind participation in the revision of this manuscript. Thanks to Dr. Raymond Tremblay for commenting on the statistical part of the manuscript.

Conflict of interest

No conflict of interest.

Author details

Michelle Jovanne Rivera-Rivera[1,2]*, Veronica Acosta-Martínez[3] and Elvira Cuevas[1,2]

1 Biology Department, University of Puerto Rico, Rio Piedras, Puerto Rico

2 Center for Applied Tropical Ecology and Conservation, University of Puerto Rico, Rio Piedras, Puerto Rico

3 United States Department of Agriculture-Agricultural Research Service, Lubbock Texas

*Address all correspondence to: yersiniamjr@hotmail.com

References

[1] Nannipieri P, Ascher J, Ceccherini MT, Landi L, Pietramellara G, Renella G. Microbial diversity and soil functions. European Journal of Soil Science. 2017;**68**(1):12-26

[2] Fierer N, Leff JW, Adams BJ, Nielsen UN, Bates ST, Lauber CL, et al. Cross-biome metagenomic analyses of soil microbial communities and their functional attributes. Proceedings of the National Academy of Sciences. 2012;**109**(52):21390-21395

[3] Van Der Heijden MG, Bardgett RD, Van Straalen NM. The unseen majority: Soil microbes as drivers of plant diversity and productivity in terrestrial ecosystems. Ecology Letters. 2008;**11**(3):296-310

[4] Acosta-Martínez V, Bell CW, Morris BE, Zak J, Allen VG. Long-term soil microbial community and enzyme activity responses to an integrated cropping-livestock system in a semi-arid region. Agriculture, Ecosystems and Environment. 2010;**137**(3-4):231-240

[5] Acosta-Martinez V, Cruz L, Sotomayor-Ramirez D, Pérez-Alegría L. Enzyme activities as affected by soil properties and land use in a tropical watershed. Applied Soil Ecology. 2007;**35**(1):35-45

[6] Parham JA, Deng SP. Detection, quantification and characterization of β-glucosaminidase activity in soil. Soil Biology and Biochemistry. 2000;**32**(8-9):1183-1190

[7] Ekenler M, Tabatabai MA. β-Glucosaminidase activity as an index of nitrogen mineralization in soils. Communications in Soil Science and Plant Analysis. 2004;**35**(7-8):1081-1094

[8] Eivazi F, Tabatabai MA. Phosphatases in soils. Soil Biology and Biochemistry. 1977;**9**(3):167-172

[9] Sardans J, Peñuelas J. Drought decreases soil enzyme activity in a Mediterranean *Quercus ilex* L. forest. Soil Biology and Biochemistry. 2005;**37**(3):455-461

[10] Castro-Díez P, Fierro-Brunnenmeister N, González-Muñoz N, Gallardo A. Effects of exotic and native tree leaf litter on soil properties of two contrasting sites in the Iberian Peninsula. Plant and Soil. 2012;**350**(1-2):179-191

[11] Wardle DA, Bardgett RD, Klironomos JN, Setälä H, Van Der Putten WH, Wall DH. Ecological linkages between aboveground and belowground biota. Science. 2004;**304**(5677):1629-1633

[12] Bruno JF, Stachowicz JJ, Bertness MD. Inclusion of facilitation into ecological theory. Trends in Ecology & Evolution. 2003;**18**(3):119-125

[13] Baldrian P, Šnajdr J, Merhautová V, Dobiášová P, Cajthaml T, Valášková V. Responses of the extracellular enzyme activities in hardwood forest to soil temperature and seasonality and the potential effects of climate change. Soil Biology and Biochemistry. 2013;**56**:60-68

[14] Medina E, Cuevas E, Molina S, Lugo AE, Ramos O. Structural variability and species diversity of a dwarf Caribbean dry forest. Caribbean Journal of Science. 2010;**46**(2-3):203-215

[15] Cintrón BB, Lugo AE. Litterfall in a subtropical dry forest: Guanica, Puerto Rico. Acta Científica. 1990;**4**(1-3):37-49

[16] Barberena-Arias MF. Single tree species effects on temperature, nutrients and arthropod diversity in litter and humus in the Guánica dry forest [Ph.D. dissertation]. University of Puerto Rico; 2008

[17] Medina E, Cuevas E. Propiedades fotosintéticas y eficiencia de uso de agua de plantas leñosas del bosque deciduo de Guánica: Consideraciones generales y resultados preliminares. Acta Científica. 1990;**4**:25-36

[18] Govender Y, Cuevas E, Sternberg LD, Jury MR. Temporal variation in stable isotopic composition of rainfall and groundwater in a tropical dry forest in the northeastern Caribbean. Earth Interactions. 2013;**17**(27):1-20

[19] Gee GW, Bauder JW. Particle-Size Analysis 1. Soil Science Society of America. Madison, Wl USA: American Society of Agronomy; 1986

[20] Anderson J, Ingram J. Tropical Soil Biology and Fertility (TSBF): Handbook of Methods. Wallingford, Ox UK: CAB International; 1993

[21] Schutter ME, Dick RP. Comparison of fatty acid methyl ester (FAME) methods for characterizing microbial communities. Soil Science Society of America Journal. 2000;**64**(5):1659-1668

[22] Cuevas E, Medina E. Nutrient dynamics within Amazonian forests. Oecologia. 1988;**76**(2):222-235

[23] Pérez-Harguindeguy N, Díaz S, Cornelissen JH, Vendramini F, Cabido M, Castellanos A. Chemistry and toughness predict leaf litter decomposition rates over a wide spectrum of functional types and taxa in Central Argentina. Plant and Soil. 2000;**218**(1-2):21-30

[24] Schneider T, Keiblinger KM, Schmid E, Sterflinger-Gleixner K, Ellersdorfer G, Roschitzki B, et al. Who is who in litter decomposition? Metaproteomics reveals major microbial players and their biogeochemical functions. The ISME Journal. 2012;**6**(9):1749

[25] Tremont O, Cuevas E. Carbono orgánico, nutrientes y cambios estacionales de la biomasa microbiana en las principales especies de dos tipos de bosques tropicales. Multiciencias. 2004;**4**(2):96-103

[26] Dijkman NA, Boschker HT, Stal LJ, Kromkamp JC. Composition and heterogeneity of the microbial community in a coastal microbial mat as revealed by the analysis of pigments and phospholipid-derived fatty acids. Journal of Sea Research. 2010;**63**(1):62-70

[27] Muyzer G, Stams AJ. The ecology and biotechnology of sulphate-reducing bacteria. Nature Reviews Microbiology. 2008;**6**(6):441

[28] Waldrop MP, Balser TC, Firestone MK. Linking microbial community composition to function in a tropical soil. Soil Biology and Biochemistry. 2000;**32**(13):1837-1846

[29] Richardson AE, Simpson RJ. Soil microorganisms mediating phosphorus availability update on microbial phosphorus. Plant Physiology. 2011;**156**(3):989-996

[30] George E, Marschner H, Jakobsen I. Role of arbuscular mycorrhizal fungi in uptake of phosphorus and nitrogen from soil. Critical Reviews in Biotechnology. 1995;**15**(3-4):257-270

[31] Wu QS, Zou YN, He XH. Differences of hyphal and soil phosphatase activities in drought-stressed mycorrhizal trifoliate orange (*Poncirus trifoliata*) seedlings. Scientia Horticulturae. 2011;**129**(2):294-298

Chapter 7

Methanogenic Diversity and Taxonomy in the Gastro Intestinal Tract of Ruminants

Farah Naz Faridi and Saba Khan

Abstract

To elucidate the microbial dynamics inside rumen of animals of livestock importance and to provide a better ration to them in order to control various metabolic disorders, a better understanding of the rumen microbial ecology is pivotal. The fundamental knowledge of methanogenic population inside gut environment and ruminal fermentation is of considerable importance as it has a significant impact on the various metabolic activities of the animal. The major methanogens isolated and characterized from ruminants like cattle, sheep, steers, goats, reindeers are from the order *Methanobacteriales*, *Methanomicrobiales*, *Methanococcales*, *Methanosarcinales* and *Methanomassiliicoccales*. The chapter deals with present knowledge available regarding the methanogenic diversity present in the gastrointestinal tract of ruminants all over the world primarily through constructing 16S rRNA gene clone libraries and tries to uncover the new genera in ruminant's microbiome and their adaptations in extreme environment. To get a better idea regarding the composition of methanogen community, further studies are required in relation to the effect of diet and animal species to the rumen methanogens.

Keywords: *Archaea*, gut, methanogens, microbiome, *Methanobrevibacter* spp., rumen

1. Introduction

The methanogens are one of the primitive life forms on earth which have evolved to be able to thrive in extreme harsh temperatures (severe hot and cold) and living conditions (salt and pH) uninhabitable for most of other life forms. Although a vast proportion of methanogens are *Archaea* but protists like algae, fungi and protozoa also form a diversity of this group. Besides their anthropogenic existence, methanogens are present in a wide area of ecological niches ranging from peat bogs to deep sea sediments and hydrothermal vents and hot springs [1, 2].

The large number of microbial population in natural anaerobic systems remains unexplored as enumeration techniques like selective enrichment, pure-culture Isolation, most-probable number estimates are time consuming and labor intensive. Culture less approaches has allowed deciphering the diversity of microbial community thriving across wide environmental ranges. Various anaerobic culture techniques led to the discovery of a third microbial kingdom, the *Archaebacteria*, which includes methanogens [3, 4]. Further the target specific sequence analysis of 16S rRNA gene in 1970's had redefined taxonomy of all living organisms into three main domains. Methanogens belong to the 3rd domain of life-*Archaea*, other two being—*Eucarya* and *Bacteria*. *Archaea* is further divided into phylums *Crenarchaeota* and *Euryarchaeota* [5].

2. Major rumen microbes

At any time there are billions of any species of anaerobic bacteria and facultative anaerobic bacteria residing in rumen along with a mixed population of various anaerobic protozoa, anaerobic fungi and flagellates making it a diverse microbial consortium in nature. The bacteria along with protozoa make most of the microbial mass (nearly 80%) inside rumen. The bacteria present in specialized niches are a very small fraction that cannot be recovered by cultural methods and even among cultivable bacteria true number of diversity is now revealed only by molecular techniques [6]. The bacteria can further be cellulolytic (fiber digesting), amylolytic (starch and sugar digesting) and lactate utilizing bacteria. The role of symbiotic microbial ecosystem consisting of bacteria, protozoa and fungi is of great significance in ruminants. Phylum Euryarchaeota within domain *Archaea* includes 7 orders—*Methanobacteriales*, *Methanomicrobiales*, *Methanococcales*, *Methanopyrales*, *Methanocellales*, *Methanosarcinales* and *Methanomassiliicoccales*. The orders are further divided into 10 families and 31 genera [7–9].

3. Methane production in ruminants and its contribution to greenhouse gases

Methane is a main byproduct of digestion in ruminants produced by the microbial fermentation of plant biomass. Methanogens ferment the ingested feed into short chain fatty acids which consists of 70% of the total metabolizable energy source for ruminants. The methane is specifically produced by methanogens (*Archaea*) that resides symbiotically in the gut of ruminants by using hydrogen produced by bacteria, fungi and protozoa and reducing CO_2 to methane. It is not used by ruminants and is lost in environment through eructation resulting in a loss of 2–12% of metabolic energy intake to the host [10, 11]. Among agricultural sources, enteric fermentation along with natural and man-made wetlands, animal wastes; paddy fields contribute to the release of major amount of methane in environment. Methane gas has a major global warming impact [12]. According to the fifth assessment report of Intergovernmental Panel on Climate Change (IPCC) published in 2014, global release of greenhouse gases from enteric fermentation grew from 1.4 to 2.1 GtCO2eq/yr between 1961 and 2010. The largest methane emission was by cattle (75% of the total) followed by goat, sheep and other ruminants during the year 2000–2010 [13]. The enteric fermentation in ruminants is a significant cause of methane emission in environment. It is an inevitable outcome of their normal digestive process [14], which is not used by them and is lost in environment. Since, methane is a potent greenhouse gas, to reduce the activity and number of methane producing *Archaea*, it is desirable to have knowledge about the community structure of methanogens and their feed conversion energy mechanism. In order to control various ruminal disorders the insight into microbial ecology will help to develop nutrition and feed management strategies.

4. Methanogenic archaeal population in gastro-intestinal tract of ruminants

The rumen was the initial environment of *Archaea* which is comprehensively investigated and studied. Hungate [15] reported that about 23 bacterial species played prominent role in ruminal metabolism whereas in 1996 the number increased up to 200 [16]. The culture based techniques had serious limitations as they failed to differentiate between two phylogenetic diverse species along with the dire need of maintaining anaerobic environment to culture and isolate bacteria. The 16S rRNA sequencing technology has been far and wide used to explore the methanogens residing inside rumen and to characterize and validate their community structure and taxonomic composition in evolutionary timeline. The methanogenic group in gastrointestinal tract of ovine, caprine and bovine using rRNA targeted oligonucleotide probes were identified and *Methanobacteriales* were reported to be the abundant methanogens in bovine and caprine rumen whereas *Methanomicrobiales* was found to be predominant in ovine rumen [17]. In 2000, the population of methanogens among rumen microbial diversity of sheep in Japan was reported using 16S rDNA cloning and fluorescence in situ hybridization (FISH) technique and most of the clones were found associated with *Methanomicrobium mobile*, *Methanobrevibacter ruminantium* and *Methanobrevibacter smithii*. The total methanogens accounted for 3.6% from the total microorganisms present in rumen and population of *M. mobile* among methanogens was found to be 54% [18]. A year later the archaeal libraries generated from the rumen of dairy Holstein cows from Japan revealed two groups of sequences produced from two different sets of archaeal primers. The library generated from primers-D30 and D33, revealed 21% of clones related to *M. mobile* and 79% of clones were anaerobic digester associated archaeal sequences with close identity to *Thermoplasma*. The second library generated from 0025e and 1492 primers showed 56% of the clones related to *M. mobile*, 20% related to the *Thermoplasma* associated sequences and 16% related to *Methanobrevibacter* spp. and 2 sequences were related to the unidentified rumen *Archaea* [19].

Similarly in bovine rumen, 41 cloned sequences were identified in 3 clusters. The largest cluster contained 24 clones with 2 distinct sub clusters with sequences affiliated with *Mbb. ruminantium*. The sub cluster Mbr I contained nine 16S rDNA sequences that had 98.5–98.8% sequence identity to *Mbb. ruminantium* whereas the sub cluster Mbr II contained 15 cloned sequences that had 97.2–97.7% similarity to *Mbb. ruminantium* whereas the second cluster contained 11 cloned sequences having similarity values of 96.1–97.5% to *Methanosphaera stadtmanae*, an organism first time recognized in rumen. The third cluster was found containing 6 cloned sequences that were 89% similar to *Methanosarcina* sp. str. WH_1 and *Methanosarcina thermophila* indicating it to be comprised of a novel group of rumen methanogens [20]. In Japan, clones were deduced from bovine rumen that was 83.9–88.3% identical to *Mbb. ruminantium* [21]. In 2004, the archaeal populations from three fractions of rumen-rumen fluid, rumen solid and rumen epithelium from Korean Hanwoo cattle was constructed using 16S rDNA gene clone libraries. Species belonging to the family *Methanomicrobiaceae* were found dominant in fractions of fluid and epithelium in rumen while *Methanobacteriaceae* was abundant in solid fraction of rumen [22]. The *Methanomicrobium* phylotype was the most abundant phylotype among methanogenic population in rumen of Murrah buffaloes from India as revealed by constructing 16S rDNA gene library. A total of 15 phylotypes out of 17 were affiliated to *M. mobile*; one sequence was identical to *T. acidophilum* and one sequence with *Methanocorpusculum bavaricum* [23]. *Methanobacteriales* was a dominant order identified from the rumen of Surti buffaloes in India by cloning and sequencing of *mcr*A gene while in an another study on Murrah buffaloes 100% sequence similarity was reported by two isolates to *Mbb. smithii* and 100% sequence similarity by one isolate to *M. mobile* based on 16S rRNA [24, 25].

5. Effect of diet on diversity of rumen methanogens

The rumen is a dynamic system therefore the microbes must change qualitatively and quantitatively in response to the changes in the chemical composition of diet of animal rather than geographical location in general. Wang et al [26] reported members of the order Rumen Cluster C (RCC) to be most abundant ruminal methanogen present in cattle from China fed agricultural residues like corn stover, rapeseed and cottonseed meals followed by the order *Methanobacteriales*. By constructing a gene clone library of *mcrA* gene, they found that by increasing the agricultural residues in diet of cattle, the methanogen community structure did not change however methane production was increased. The effect of diet on rumen methanogen population has also been studied in Western Australia where sheep were fed different diets. Analysis revealed that archaeal diversity in sheep from grazing pasture was more as compared to sheep fed forage diets-oaten hay or lucerne hay. The maximum numbers of clones identified were from *Methanobrevibacter* strains SM9, M6, and NT7 [27].

A corn and cottonseed diet of cattle from Jinnan region of China also reported members of *Methanobrevibacter, Methanobacterium*, *Methanosphaera, Methanomicrobium* and unidentified Euryarchaeota. Overall, *Methanobrevibacter* spp. appeared to be predominant in all three rumen fractions [28]. Similarly, methanogenic population in dairy cattle from Canada was estimated that were fed diets supplemented with enzyme additive by PCR-DGGE and quantitative real-time PCR (qRT-PCR) analysis. The PCR-DGGE profiles were made up of 26 different bands, with two bands affiliated to Methanogenic archaeon CH1270 and one band to *Mbb. gottschalkii* strain HO. Three bands similar to Methanogenic archaeon CH1270 or *Mbb. smithii* ATCC 35061 appeared after enzyme was supplemented [29]. The diversity of rumen methanogens present in Mediterranean water buffaloes from Brazil which were maintained on three different diets-corn silage (library 1), pasture grazing (library 2) and sugar cane (library 3) revealed all three 16S rRNA clone libraries to be consisted of *Methanobrevibacter*-related sequences. The abundance of *Methanobrevibacter* like sequences in water buffaloes was in contrast to previous reports that showed *M. mobile* like methanogens to be predominant *Archaea* isolated from water buffaloes of Murrah and Surti breeds from India [30]. The taxonomy and structure of methanogens in Swedish dairy cattle fed two different diets through clone library consisted by terminal restriction fragment length polymorphism (T-RFLP) showed the genus *Methanobrevibacter* to be dominant in rumen and that the diet may not be an obvious factor affecting the community composition of methanogenic population inside rumen but may give an insight to the structure of ruminal methanogens [31].

Another study on sheep in Queensland, Australia in 2006 revealed 78 clones of 26 different methanogen related sequences were obtained. Eight sequences consisted of 15 clones were found 95–100% similar to the orders *Methanobacteriales* and *Methanomicrobiales*, and rest 18 sequences consisted of 63 clones were 72–75% affiliated to *Thermoplasma acidophilum* (*T. acidophilum*) and *Thermoplasma volcanium* (*T. volcanium*) [32]. The structure of archaeal diversity in feedlot cattle (starch based diet) from two different provinces of Canada-Ontario and Prince Edward Island, were deduced by constructing a clone library of 241 sequences.

Eleven phylotypes (38 clones) in cattle from Ontario region (corn-based diet) were unique to this group as they were not found in cattle from Prince Edward Island. Similarly, 7 phylotypes (42 clones) from Prince Edward Island cattle (potato by-products) were found only in this group whereas 5 sequences representing 161 clones were found common in both herds. Out of 23 different sequences obtained, 10 sequences consisting of 136 clones were 89.8–100% affiliated to the species of the orders *Methanobacteriales*, *Methanomicrobiales* and *Methanosarcinales* and remaining 13 sequences consisting of 105 clones showed 74.1–75.8% sequence similarity to the species *T. volcanium* and *T. acidophilum* [33]. The dominance of total rumen *Archaea* from different ruminant species around the world in a global data set report surveying nine studies assessed that genus *Methanobrevibacter* (61.6%), *Methanomicrobium* (14.9%) and uncultured species from Rumen Cluster C (15.8%) constituted 92.3% of total rumen *Archaea* [34]. Another study from Venezuela indicated *Methanobrevibacter* phylotype to be the most abundant genera in 14 different 16S rRNA gene sequences or phylotypes from 104 clone library constructed in sheep [35]. The rumen of Sika deer fed oak leaf diet and corn stalk diet from China revealed thirty six OTUs assigned to 146 unique sequences and in both the diet group, genus *Methanobrevibacter* was detected as a predominant methanogen. Among the species, *Mbb. millerae* was most abundant in both groups but accounted for a slightly higher population (69.5%) in corn stalk library than in oak leaf library (51.4%). Clones with similarity to *Mbb. smithii* like clones and *Mbb. ruminantium* like clones were present in corn stalk library but were absent in oak leaf library [36].

The majority of sequences were related to genera *Methanobrevibacter* and *Methanosphaera* and a group of novel uncultured methanogens "uncultured marine bacteria" were identified in Moxoto breed goats from Brazil by constructing 16S rRNA gene clone libraries [37]. Likewise, the archaeal methanogen population inside rumen of lactating Jersey and Holstein cattle fed same diet from America revealed species level similarity to *Mbb. ruminantium* [38]. The community structure of methanogens inside rumen of farmed sheep, cattle and red deer which were fed different diets revealed diet and host based differences in framing community structure, but the presence of dominant archaeal species was uniform in all host animals. The dominant members were from following clades: RO clade-*Mbb. ruminantium* and *Mbb. olleyae*, SGMT clade-*Mbb. gottschalkii, Mbb. millerae and Mbb. thaueri* and species of the genus *Methanosphaera* [39].

6. Methanogen diversity inside rumen and/feces under similar conditions of diet

The sequences obtained from rumen and feces of local sheep from Xinjiang, China were divided into three groups based on their affiliation to the following genera: *Methanobrevibacter*, *Methanocorpusculum* and an unclassified methanogen-like group [40]. Order *Methanobacteriales* was found to be dominant in rumen of faunated and unfaunated Holstein cattle from Japan by constructing clone libraries from 16S rDNA gene and mcrA gene [41]. The methanogenic archaeal population in sheep of Scottish uplands were illustrated by Snelling et al. [42] by different methods-Sanger amplicon sequencing by constructing 16S rRNA gene libraries, 16S rRNA gene amplicon sequencing by Illumina, Illumina metagenome sequencing. All the methods revealed the order *Methanobacteriales* containing genera: *Methanobrevibacter*, *Methanosphaera* and *Methanobacteria* to be the most abundant. Among the *Methanobacteriales* order, *Mbb. millerae* comprised of ≥91% of OTU's and remainder of the OTU's were formed by *Methanosphaera*.

Tymensen and McAllister [43] reported the archaeal spp. linked with ruminal protozoa in cattle and obtained 276 final sequences generated from clone libraries using five diverse universal archaeal primer pairs and found that the three genera/taxa viz. *Methanobrevibacter*, Rumen Cluster C (RCC) and *Methanomicrobium* accounted for 94–100% of the sequences in each library. Metatranscriptomics approach-Illumina deep-sequencing with overlapping read paired-end technology revealed that *Bacteria* and *Eukaryotes* contributed to the majority of ribotags (approximately 50%) whereas *Archaea* contributed only 1% of ribotags mainly comprised of the order *Methanobacteriales* (*Methanobrevibacter* and *Methanosphaera*) and *RCC Thermoplasmata*. The *RCC Thermoplasmata* lowered down considerably on rape seed oil (RSO) supplementation whereas *Methanobacteriales* did not show any decrease. A notable decrease in the mcrA and mcrB transcripts of RCC on change in was noticed suggesting the reduced CH_4 emissions [44].

The abundance of two archeal orders-*Methanobacteriales* and *Methanomassiliicoccales* in rumen of sheep and cattle from New Zealand were studied. From the order *Methanobacteriales*, sequences were assigned to only four species—*Mbb. gottschalkii*, *Mbb. ruminantium*, *Methanosphaera* sp. ISO3-F5 and *Methanosphaera* sp. group5. The members of the order *Methanomassiliicoccales* contributed 10.4% of the total relative abundance of the methanogenic archaeal community, *Methanobacteriales* (89.6%) being dominant [45]. The methanogenic *Archaea* in yak from China grazing on natural pastures exhibited the species of the family *Methanobacteriaceae* to be predominant in yak rumen followed by members from the family *Methanomassiliicoccaceae* and *Methanosarcinaceae* [46].

The archaeal methanogenic community from rumen of two indigenous ruminant species-yak and Tibetan sheep and two introduced species-cattle and crossbred sheep in Qinghai-Tibetan plateau from China under similar diet of oaten hay and barley and environmental conditions revealed the more archaeal diversity in indigenous species than in introduced species. *Methanomassiliicoccaceae* was predominant family representing most of the sequences while *Methanobacteriaceae* was second most dominant archaeal family. Among *Methanobrevibacter* genus, *Mbb. gottschalkii* and *Mbb. ruminantium* were the most observed species. Interestingly, *Mbb. woesei* and *Mbb.* sp. RT were only found associated with yak rumen [47]. Salgado-Flores et al. [48] reported archaeal methanogenic density by quantitative real-time PCR and diversity from rumen and cecum samples of Norwegian reindeer fed on standard pellets and lichens by 454 pyrosequencing of 16S rRNA genes. The population density of archaeal methanogens remained almost constant for both the diets in rumen and cecum samples. In rumen samples, *Methanobrevibacter* was found to be main genus and strain *Mbb. thaueri* CW to be predominant in both groups fed different diets. *Mbb. wolinii* SH was second most abundant species found in group fed pellet based diet whereas constituted only 1.5% of the total sequences in group fed lichens. The second most prevalent species was *Mbb. ruminantium* strain M1 in reindeers fed lichens but accounted only 4.2% of the total sequences in pellet fed group of reindeers. In cecum samples also, genus *Methanobrevibacter* was detected predominantly in both the groups. *Mbb. millerae* strain ZA-10 was found to be most abundant in reindeer group fed with pellet but had less than 97% similarity with this archaeal methanogen whereas strain *Mbb. thaueri* CW was main species in lichen fed group with 98% similarity. Franzolin and Wright reported that the density of archaeal methanogens was very low as compared to bacterial counterparts in grazing and feedlot group of buffaloes from Brazil. The density of methanogens as compared to bacteria in reticulum was more as compared to rumen [49].

The rumen methanogenic structure in three Indian cattle and buffaloes which were fed on wheat straws based diet using RT-PCR revealed most abundant orders of *Methanomicrobiales* and *Methanobacteriales* along with total bacteria and that it remained constant for two animals using a particular diet [50]. Similarly, the ruminal diversity in Indian Murrah buffaloes by using amplified ribosomal DNA restriction analysis (ARDRA) maintained under standard diet of wheat straws revealed a total of 108 clones that were classified into 16 phylotypes. The 9 phylotypes showed less than 97% sequence similarity to any of the cultivated methanogen strain and represented a novel uncultured group of methanogens. The second group comprised of 4 phylotypes that showed 92–99% sequence similarity with *M. mobile*. The third group consisted of a single phylotype clustered with *M. burtonii*, reported for

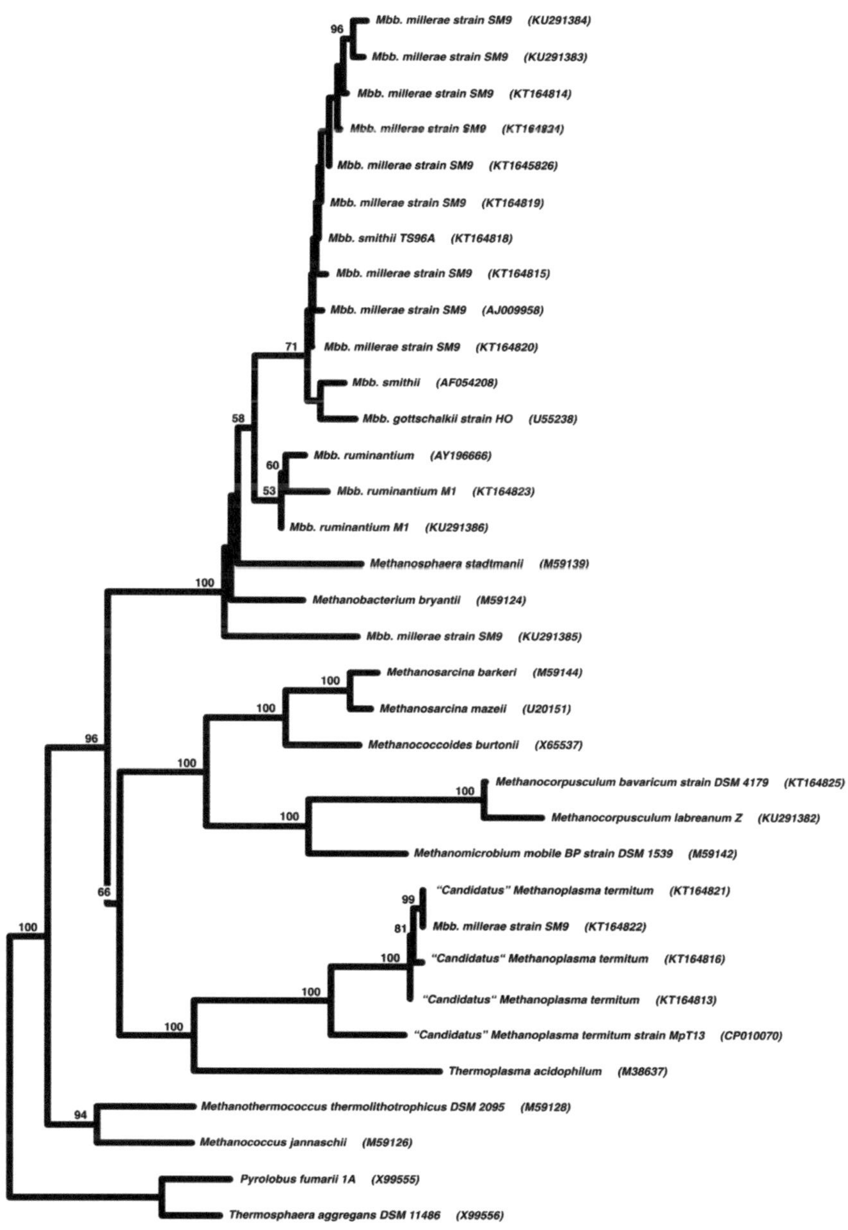

Figure 1.
A phylogenetic tree based on 16S rRNA sequences obtained from camel foregut and reference sequences downloaded from NCBI Genbank database [58].

the first time in rumen. The fourth group was a single phylotype that showed 97% sequence identity with *Mbb. gottschalkii*. The last group of single phylotype showed a sequence similarity to *Mbb. ruminantium* [51].

Likewise, the comparative diversity analysis of methanogens using 16S rRNA and *mcr*A in cattle rumen fed on a high fiber diet reported 13 OTU's consisting of 102 clones from 16S rRNA gene based library. All OTU's were clustered with order *Methanobacteriales* and were further splitted into Cluster I that had 12 OTU's related to *Methanobrevibacter* spp. and Cluster II comprised of one OTU related to *M. stadtmanae* [52]. The Surti buffaloes that were fed wheat straw and compound concentrate mixture diet generated a total of 76 clones representing 21 sequences based on PCR-RFLP patterns. BLAST analysis revealed 13 OTU's (55 clones) that showed sequence identity with *Methanomicrobium* sp., 3 OTU's (15 clones) that showed sequence similarity with *Methanobrevibacter* sp. The remaining 5 OTU's (6 clones) were associated with uncultured *Archaea*. Overall, the methanogenic population inside rumen of buffaloes was from the order of *Methanomicrobiales* (18 OTUs) and *Methanobacteriales* (3 OTUs) [53]. The rumen metagenome of buffalo using q-PCR were compared with MG-RAST based annotation of the metagenomes sequences of 16S rDNA amplicons and high throughput shotgun sequencing and found *Methanomicrobiales* in lower number [54] (**Figure 1**).

7. Methanogenic archaeal population in pseudo ruminants like camelids

Gut methanogens remains largely uncharacterized in camel with no published studies on methanogenic archaeal populations from 16S rRNA gene clone libraries whereas much interest has been paid to domestic ruminants. The community diversity and structure of archaeal methanogens in fecal samples of Bactrian camel (*Camelus bactrianus*) maintained at two zoos from United States of America revealed the genus *Methanobrevibacter* to be the abundant ruminal methanogen however the diversity and structure of methanogens varied significantly between the two libraries with only 2 OTU's in common to both the libraries. Two and seven OTU's were found unique to first and second library, respectively [55]. The methanogenic archaeal population inside rumen of Alpaca (*Vicugna pacos*) from America resulted in a 947 non chimeric gene clone library representing 51 distinct OTU's. Thirty seven OTU's displayed ≥95% genus-level sequence affiliation with the species belonging to *Methanobrevibacter*. Six out of 37 OTU's showed ≥98% species-level sequence identity to *Mbb. millerae*; 2 OTU's showed species-level identity to *Mbb. ruminantium*; 2 OTU's showed >98% identity to *Mbb. smithii*; 27 OTU's showed 95–97.9% sequence similarity to well detected and reported *Methanobrevibacter* species. Of the remaining 14 OTU's, 3 distinct phylogenetic group were made that consisted of 4 OTU's that had 95–97.9% similarity to the species of *Methanobacterium*; other 3 OTU's showed genus level similarity with the species *of Methanosphaera*; 7 OTU's were found to be isolated phylogenetically from order *Methanobacteriales*. Overall, *Methanobrevibacter* was found to be dominant in alpaca rumen like other ruminants but in contrast as described in other ruminants *Mbb. millerae* was found to be in most number of clones showing species level identity [56]. The fecal microbiome of camels maintained at intensive and extensive system of management in Jaisalmer (Rajasthan) was evaluated through non-cultural approach. The both group's fecal metagenomes were compared with available fecal or rumen metagenomes on MG-RAST and *Mbb. smithii* was detected as a predominant archaeal methanogen [57]. A 16S rRNA gene clone library from the content of the C1 compartment (foregut) of Indian camels was constructed by cloning pooled polymerase chain reaction (PCR)—amplified products. The sequences (n = 151) were clustered into 15 OTU's (operational taxonomic units) based on sequencing of unique RFLP pattern and divided into five species groups: *Methanobrevibacter* (*Mbb.*) *millerae* strain SM9, "*Candidatus*" Methanoplasma termitum, *Mbb. smithii*, *Mbb. ruminantium*, *Methanocorpusculum* (*M.*) *bavaricum* strain DSM 4179. The genus *Methanobrevibacter* (order *Methanobacteriales*) was the most prevalent (76.82%), followed by *Archaea* from the orders *Methanomassiliicoccales* (17.21%) and *Methanomicrobiales* (5.96%) [58] (**Figure 1**).

8. Biotechnological applications of extremophiles

The microbial diversity of extremophiles is of interest particularly for microbiologists and biotechnologists to decipher the enzymes and their functions, their biochemical and metabolic pathways that enable them to survive in harshest conditions. The in depth knowledge will pave the path for creating technologies that can function under extreme conditions. It will improve our current knowledge and perception about the interrelationships between various species and will continue to lead to the classification and assessment of ruminal archaeal species.

For researchers working to explore the microbial ecology of volcanic systems, deep under the earth, oceans, thermal vents, rice fields, waste treatments, bioremediation of soils, the rumen forms a stable and basic source of knowledge concerning anaerobic microorganisms. The knowledge of anaerobic microorganism's reaction going inside rumen flora is of invaluable importance as methanogens are also found in omnivores and humans alike and can be implicated in understanding human and animal diseases. An extensive understanding of methanogens in gastrointestinal tract will contribute to the sustainable farming of animals well into the future. The enteric fermentation in ruminants is a significant cause of methane emission in environment. Since, methane is a potent greenhouse gas, to reduce the activity and number of methane producing *Archaea*, it is desirable to have knowledge about the community structure of methanogens and their feed conversion energy mechanism. In order to control various ruminal disorders the insight into microbial ecology will help to develop nutrition and feed management strategies and also to develop better prospects of altering rumen function to mitigate methane generation while still optimizing digestibility and microbial function. This can be particularly useful for the farmer community who can benefit environment in methane mitigation from livestock at the same time increasing animal efficiency. Reductive acetogenesis is performed by acetogenic bacteria that thrive in non-ruminants and can sometimes replace methanogenesis. A comparative account of dominant methanogens in the ruminants all over the world is depicted in **Figure 2**.

The significance of exploring the archaeal diversity lies in its great potential to identify the genes encoding plant degrading enzymes, thus contributing to an increase in understanding of the mechanisms mediating digestion in ruminants. Moreover, the functional analysis of these genes might uncover strategies for improving feed and fiber digestion in the rumen that could further be applied to manipulate pathways associated with bioreactor processes for biofuels production and to formulate feed with dietary additives that help in reducing methane emissions. A taxonomic frame of methanogens should be developed that would help elucidate the diversity, identification and classification of major rumen archaeal population. Data from antibiotic resistance genes and RATC (resistance to antibiotics and toxic compounds) can be also used to produce antibiotic resistance gene profiles to help in understanding of the microbial community ecology in every environment.

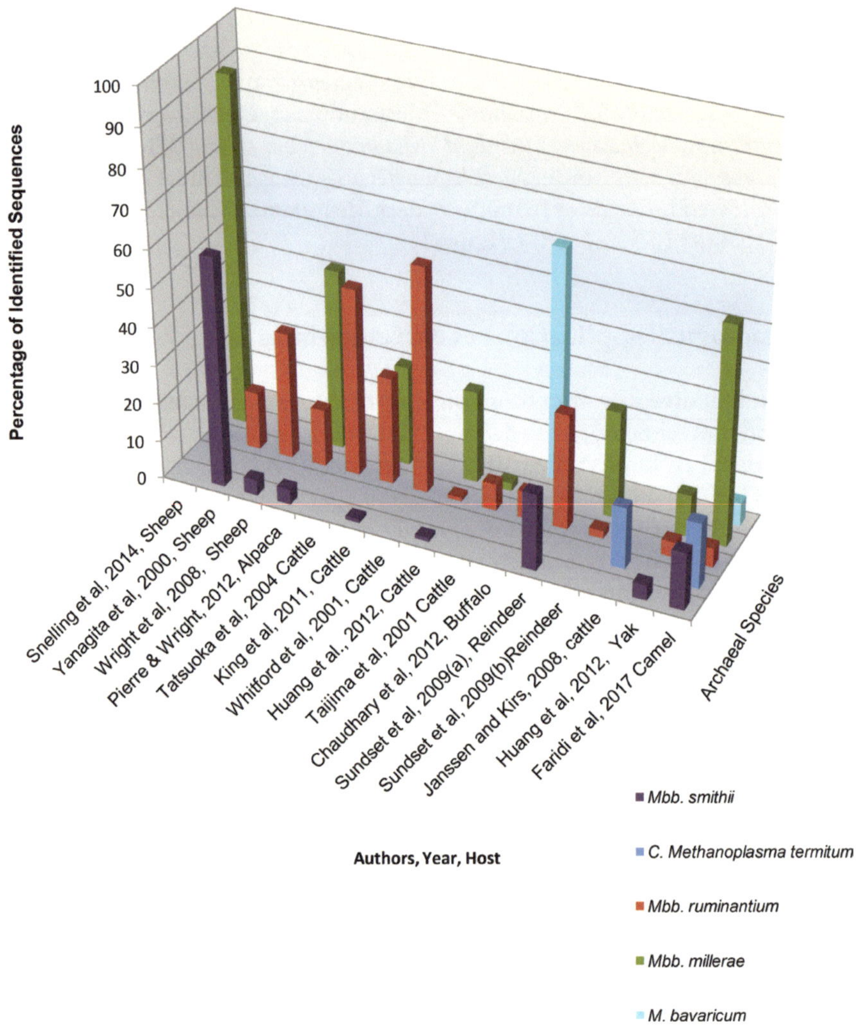

Figure 2.
Methanogenic sequences identified in ruminants around the world.

One can exploit enzymes from extremophile *Archaea* that can endure high temperatures and organic solvents. Acidophiles are used in coal mining to recover metallic minerals and to reduce sulfur levels. Alkaliphiles are used in paper making and spilled oil recovery, besides being used as a common ingredient in dish washing detergent and laundry soap. *Thermus aquaticus* an extremophile that endures high temperature produces an enzyme called *Taq* polymerase that has transformed molecular biology all over the world by aiding in quick DNA replication during polymerase chain reaction (PCR). The extremophiles are immensely used in medical and food microbiology, industrial fermentations to produce acetone, butanol, etc. The understanding of microbial diversity in extreme habitats like wetlands can propose research strategies and priorities to integrate understanding of plant-microbial interactions. Further, studies should provide the break through to link distribution and distinctiveness of various gastrointestinal microbes in their natural environment and to discover their genetic potential for livestock wellbeing and industrial progress by making a significant contribution in understanding ruminant nutrition. Research in microbial genomics will provide the opportunity to make sure that this knowledge is used to enhance ruminant production through an improved understanding of microbial function and ecology.

Author details

Farah Naz Faridi[1]* and Saba Khan[2]

1 Department of Bioscience and Biotechnology, Banasthali Vidyapith, Tonk, Rajasthan, India

2 Department of Veterinary Pathology, Post Graduate Institute of Veterinary Education and Research, Jaipur, Rajasthan, India

*Address all correspondence to: farah.faridi@gmail.com

References

[1] Takai K, Horikoshi K. Genetic diversity of *Archaea* in deep-sea hydrothermal vent environments. Genetics. 1999;**152**(4):1285-1297

[2] Florin TH, Zhu G, Kirk KM, Martin MG. Shared and unique environmental factors determine the ecology of methanogens in humans and rats. The American Journal of Gastroenterology. 2000;**95**(10):2872-2879

[3] Hungate RE. The Rumen and its Microbes. New York: Academic Press Inc; 2006

[4] Bryant MP. Bacterial species of the rumen. Bacteriological Reviews. 1959;**23**(3):125-153

[5] Woese CR, Kandler O, Wheelis ML. Towards a natural system of organisms: Proposal for the domains *Archaea*, Bacteria, and Eucarya. Proceedings of the National Academy of Sciences of the United States of America. 1990;**87**(12):4576-4579

[6] Hobson PN, Stewart CS. The Rumen Microbial Ecosystem. London: Chapmann & Hall; 1997

[7] Liu Y, Whitmann WB. Metabolic, phylogenetic, and ecological diversity of the methanogenic *Archaea*. Annals of the New York Academy of Sciences. 2008;**1125**:171-189

[8] Sakai S, Imachi H, Hanada S, Ohashi A, Harada H, Kamagata Y. *Methanocella paludicola* gen. nov., sp. nov., a methane-producing archaeon, the first isolate of the lineage 'Rice Cluster I', and proposal of the new archaeal order Methanocellales ord. nov. International Journal of Systematic and Evolutionary Microbiology. 2008;**58**(4):929-936

[9] Paul K, Nonoh JO, Mikulski L, Brune A. *"Methanoplasmatales,"* *Thermoplasmatales*-related *Archaea* in termite guts and other environments, are the seventh order of methanogens. Applied and Environmental Microbiology. 2012;**78**(23):8245-8253

[10] Johnson KA, Johnson DE. Methane emissions from cattle. Journal of Animal Science. 1995;**73**(8):2483-2492

[11] Kumar S, Puniya AK, Puniya M, Dagar SS, Sirohi SK, Singh K, et al. Factors affecting rumen methanogens and methane mitigation strategies. World Journal of Microbiology and Biotechnology. 2009;**25**(9):1557-1566

[12] Broucek J. Production of methane emissions from ruminant husbandry: A review. Journal of Environmental Protection. 2014;**5**:1482-1493

[13] FAO. FAOSTAT Database Collections. Food and Agriculture Organization of the United Nations; 2014. Available from: http://faostat.fao.org/

[14] Moss AR, Jouany JP, Newbold J. Methane production by ruminants: Its contribution to global warming. Annales de Zootechnie. 2000; **49**:231-253

[15] Hungate RE. Symposium: Selected topics in microbial ecology. Microbial ecology of the rumen. Bacteriological Reviews. 1960;**24**(4):353-364

[16] Krause DO, Russell JB. An rRNA approach for assessing the role of obligate amino acid-fermenting bacteria in ruminal amino acid deamination. Applied and Environmental Microbiology. 1996;**62**(3):815-821

[17] Lin C, Raskin L, Stahl DA. Microbial community structure in gastro intestinal tracts of domestic animals: Comparative analyses using rRNA targeted oligonucleotide probes. FEMS Microbiology Ecology. 1997;**22**:281-294

[18] Yanagita K, Kamagata Y, Kawaharasaki M, Suzuki T, Nakamura Y, Minato H. Phylogenetic analysis of methanogens in sheep rumen ecosystem and detection of *Methanomicrobium mobile* by flourescence *in situ* hybridization. Bioscience, Biotechnology, and Biochemistry. 2000;**64**(8):1737-1742

[19] Tajima K, Nagamine T, Matsui H, Nakamura M, Aminov RI. Phylogenetic analysis of archaeal 16S rRNA libraries from the rumen suggests the existence of a novel group of *Archaea* not associated with known methanogens. FEMS Microbiology Letters. 2001;**200**(1):67-72

[20] Whitford MF, Teather RM, Forster RJ. Phylogenetic analysis of methanogens from the bovine rumen. BMC Microbiology. 2001;**1**(1):5

[21] Tatsuoka N, Mohammed N, Mitsumori M, Hara K, Kurihara M, Itabashi H. Phylogenetic analysis of methyl coenzyme-M reductase detected from the bovine rumen. Letters in Applied Microbiology. 2004;**39**(3):257-260

[22] Shin EC, Choi BR, Lim WJ, Hong SY, An CL, Cho KM, et al. Phylogenetic analysis of *Archaea* in three fractions of cow rumen based on the 16S rDNA sequence. Anaerobe. 2004;**10**(6):313-319

[23] Chaudhary PP, Sirohi SK. Dominance of *Methanomicrobium* phylotype in methanogen population present in Murrah buffaloes (*Bubalus bubalis*). Letters in Applied Microbiology. 2009;**49**:274-277

[24] Singh KM, Pandya PR, Parnekar S, Tripathi AK, Ramani U, Koringa PG, et al. Methanogenic diversity studies within the rumen of Surti buffaloes based on methyl coenzyme M reductase A (*mcrA*) genes point to *Methanobacteriales*. Polish Journal of Microbiology. 2010;**59**(3):175-178

[25] Kumar S, Dagar SS, Puniya AK. Isolation and characterization of methanogens from rumen of Murrah buffalo. Annales de Microbiologie. 2011;**62**(1):345-350

[26] Wang P, Zhao S, Wang X, Zhang Y, Zheng N, Wang J. Ruminal methanogen community in dairy cows fed agricultural residues of corn stover, rapeseed, and cottonseed meals. Journal of Agricultural and Food Chemistry. 2016;**64**(27):5439-5445

[27] Wright A-DG, Williams AJ, Winder B, Christophersen CT, Rodgers SL, Smith KD. Molecular diversity of rumen methanogens from sheep in Western Australia. Applied and Environmental Microbiology. 2004;**70**(3):1263-1270

[28] Pei CX, Mao SY, Cheng YF, Zhu WY. Diversity, abundance and novel16S rRNA gene sequences of methanogens in rumen liquid, solid and epithelium fractions of Jinnan cattle. Animal. 2010;**4**(1):20-29

[29] Zhou M, Chung YH, Beauchemin KA, Holtshausen L, Oba M, McAllister TA, et al. Relationship between rumen methanogens and methane production in dairy cows fed diets supplemented with a feed enzyme additive. Journal of Applied Microbiology. 2011;**111**(5):1148-1158

[30] Franzolin R, St-Pierre B, Northwood K. Analysis of rumen methanogen diversity in water buffaloes (*Bubalus bubalis*) under three different diets. Microbial Ecology. 2012;**64**(1):131-139

[31] Danielsson R, Schnürer A, Arthurson V, Bertilsson J. Methanogenic population and CH_4 production in Swedish dairy cows fed different levels of forage. Applied and Environmental Microbiology. 2012;**78**(17):6172-6179

[32] Wright A-DG, Toovey AF, Pimm C. Molecular identification of methanogenic *Archaea* from sheep in Queensland, Australia reveal more uncultured novel *Archaea*. Anaerobe. 2006;**12**(3):134-139

[33] Wright A-DG, Auckland C, Lynn DH. Molecular diversity of methanogens in feedlot cattle from Ontario and Prince Edward Island, Canada. Applied and Environmental Microbiology. 2007;**73**(13):4206-4210

[34] Jannsen PH, Kirs M. Structure of the archaeal community of the rumen. Applied and Environmental Microbiology. 2008;**74**(12):3619-3625

[35] Wright A-DG, Ma X, Obispo NE. *Methanobrevibacter* phylotypes are the dominant methanogens in sheep from Venezuela. Microbial Ecology. 2008;**56**(2):390-394

[36] Li ZP, Liu HL, Jin CA, Cui XZ, Jing Y, Yang FH, et al. Differences in the methanogen population exist in Sika deer (*Cervus nippon*) fed different diets in China. Microbial Ecology. 2013;**66**(4):879-888

[37] Cunha IS, Barreto CC, Costa OYA, Bomfim MA, Castro AP, Kruger RH, et al. Bacteria and *Archaea* community structure in the rumen microbiome of goats (*Capra hircus*) from the semiarid region of Brazil. Anaerobe. 2011;**17**(3):118-124

[38] King EE, Smith RP, St-Pierre B, A-DG W. Differences in the rumen methanogen populations of lactating Jersey and Holstein dairy cows under the same diet regimen. Applied and Environmental Microbiology. 2011;**77**(16):5682-5687

[39] Jeyanathan J, Kirs M, Ronimus RS, Hoskin SO, Janssen PH. Methanogen community structure in the rumens of farmed sheep, cattle and red deer fed different diets. FEMS Microbiology Ecology. 2011;**76**(2):311-326

[40] Liu C, Zhu ZP, Liu YF, Guo TJ, Dong HM. Diversity and abundance of the rumen and fecal methanogens in Altay sheep native to Xinjiang and the influence of diversity on methane emissions. Archives of Microbiology. 2012;**194**:353-361

[41] Ozutsumi Y, Tajima K, Takenaka A, Itabashi H. *Mcr*A gene and 16S rRNA gene in the phylogenetic analysis of methanogens in the rumen of faunated and unfaunated cattle. Animal Science Journal. 2012;**83**:727-734

[42] Snelling TJ, Genc B, McKain N, Watson M, Waters SM, Creevey CJ, et al. Diversity and community composition of methanogenic *Archaea* in the rumen of Scottish upland sheep assessed by different methods. PLoS One. 2012;**9**(9):1-9

[43] Tymensen LD, McAllister TA. Community structure analysis of methanogens associated with rumen protozoa reveals bias in universal archaeal primers. Applied and Environmental Microbiology. 2012;**78**(11):4051-4056

[44] Poulsen M, Schwab C, Jensen BB, Engberg RM, Spang A, Canibe N, et al. Methylotrophic methanogenic Thermoplasmata implicated in reduced methane emissions from bovine rumen. Nature Communications. 2013;**4**:1428

[45] Seedorf H, Kittelmann S, Janssen PH. Few highly abundant operational taxonomic units dominate within rumen methanogenic archaeal species in New Zealand sheep and cattle. Applied and Environmental Microbiology. 2015;**81**(3):986-995

[46] Xue D, Chen H, Chen F, He Y, Zhao C, Zhu D, et al. Analysis of the rumen bacteria and methanogenic *Archaea* of

yak (*Bos grunniens*) steers grazing on the Qinghai-Tibetan plateau. Livestock Science. 2016;**188**:61-71

[47] Huang XD, Martinez-Fernandez G, Padmanabha J, Long R, Denman SE, McSweeney CS. Methanogen diversity in indigenous and introduced ruminant species on the Tibetan plateau. Archaea. 2016:1-11

[48] Salgado-Flores A, Hagen LH, Ishaq SL, Zamanzadeh M, Wright A-DG, Pope PB, et al. Rumen and cecum microbiomes in reindeer (*Rangifer tarandus tarandus*) are changed in response to a lichen diet and may affect enteric methane emissions. PLoS One. 2016;**11**(5):1-22

[49] Franzolin R, A-DG W. Microorganisms in the rumen and reticulum of buffalo (*Bubalus bubalis*) fed two different feeding systems. BMC Research Notes. 2016;**9**:243

[50] Chaudhary PP, Dagar SS, Sirohi SK. Comparative quantification of major rumen microbial population in Indian cattle and Buffalo fed on wheat straws based diet. Prime Journal Microbiology Research. 2012;**2**(3):105-108

[51] Chaudhary PP, Sirohi SK, Saxena J. Diversity analysis of methanogens in rumen of *Bubalus bubalis* by 16S riboprinting and sequence analysis. Gene. 2012;**493**:13-17

[52] Sirohi SK, Chaudhary PP, Singh N, Singh D, Puniya AK. The 16S rRNA and mcrA gene based comparative diversity of methanogens in cattle fed on high fibre based diet. Gene. 2013;**523**:161-166

[53] Singh KM, Tripathi AK, Pandya PR, Parnerkar S, Kothari RK, Joshi CG. Molecular genetic diversity and quantitation of methanogen in Ruminal fluid of Buffalo (*Bubalus bubalis*) fed ration (wheat straw and concentrate mixture diet). Genetics Research International. 2013;**2013**:1-8

[54] Nathani NM, Patel AK, Dhamannapatil PS, Kothari RK, Singh KM, Joshi CG. Comparative evaluation of rumen metagenome community using qPCR and MG-RAST. AMB Express. 2013;**3**:55

[55] Turnbull KL, Smith RP, St-Pierre B, Wright A-DG. Molecular diversity of methanogens in fecal samples from Bactrian camels (*Camelus bactrianus*) at two zoos. Research in Veterinary Science. 2012;**93**(1):246-249

[56] St-Pierre B, A-DG W. Molecular analysis of methanogenic *Archaea* in the forestomach of the alpaca (*Vicugna pacos*). BMC Microbiology. 2012;**12**(1)

[57] Dande SS, Bhatt VD, Patil NV, Joshi CG. The camel faecal metagenome under different systems of management: Phylogenetic and gene-centric approach. Livestock Science. 2015;**178**:108-118

[58] Faridi F, Sena DS, Sharma V. Characterization of the methanogenic archaeal community in the C1 compartment of the camel (*camelus dromedarius*). Journal of Camel Practice and Research. 2017;**24**(2):157-163

Chapter 8

A Review on Electro-Rheological Fluid (ER) and Its Various Technological Applications

Sudipto Datta, Ranjit Barua and Jonali Das

Abstract

The technology of electro-rheological fluids is old, but now it is being used in market at high speed. In the near future, ER fluids may be used for structure design where accuracy, density and power density are primarily the main criteria. For minimizing the costs and maximizing the functionality of the products, like where the viscosity of the fluid is varied to design, solid structure ER fluids are used. Features like fast response and easy interface between electrical and mechanical input–output makes the ER fluid attractive to various technology fields. In this study, ER fluids' working principle, various low-cost ER fluids working procedures and ER fluid applications in multiple areas are explained.

Keywords: ER clutch, mechatronics, ER fluids, ER damper

1. Introduction

In electro rheological (ER fluids) the additive particles are kept in suspension in a dielectric fluid which is non-conducting. The Dielectric fluid, i.e., the Carrier fluid has high electrical resistivity and has a low viscosity like silicon oil, olive oil, hydrocarbons, etc. The additive particles which are mixed in the carrier fluids are mainly polymers, alumina silicates, metal oxides silica, etc. These additive particles commonly have low particles size which allows the carrier fluid to maintain low viscosity when the external electric field is not applied. In ER fluid the additive particles size range remains in 0.1–100 μm in the carrier fluid. Without any external electric field these fluids stays in liquid condition as soon as the external electric field is applied the ER fluid changes from liquid to solid by viscosity change of the fluid. In Electro rheological (ER) fluids a suspension of particles are present in a non-conducting fluid. The commonly used liquid i.e. hydrocarbon or silicon oil for suspension are low viscous and have high resistivity. Suspension particles are mainly polymers, alumina, silicates, metal oxides etc. These particles are present is very low concentration so that the viscosity of the suspending fluid remains low without application of the applied electric field. The suspension particles are dielectrics of size 0.1–100 μm. In absence of the electric field the particles exhibits properties like fluid and as the electric field is applied the particles behaves like solid. These fluids which change its physical properties like viscosity due to application of electric field are called electro rheological (ER) fluids or smart fluids. Types of ER or Smart fluids: (a) Electro Rheological (ER) Fluids—electric field changes the physical properties of the fluid,

(b) Magneto Rheological (MR) Fluids—magnetic fields changes the physical properties of the fluid, (c) Positive Electro Rheological (ER) Fluids—by application of the electric field the viscosity increases and (d) Negative Electro Rheological studied by Ko et al. [1] (ER) Fluids—by application of electric field the viscosity decreases. These ER fluids are one kind of smarts fluids. One of the most easily made ER fluid is adding corn flour in silicon oil or vegetable oil.

2. Electro rheological (ER) material interaction

When the electric field is applied on the ER fluid the suspension particles gets polarized and form a thick chain which is parallel to the electric field between the two electrodes. The thickness of the polarized suspension particles between the two electrodes is directly proportional to the intensity of the electric field. The rheological properties of the suspension depend on its change in structure. The more yield stress of the fluid is obtained from the particle columnar structure. When the electric field is removed the suspension particles polarization gets lost and the loose there structure and roam freely in the fluid which in turn reduces the viscosity. The period of returning from the solid state to the liquid state is few milliseconds upon removing the electric field. The material for electrorheological fluid is a superfine suspension of dielectric small particles which react to the applied electric field resulting in changing in the rheological properties of the ER fluid. There are three operational modes of the ER fluid which are as follows: (a) Flow mode—in this mode the electrodes are mounted and fixed and by controlling the motion of the flow the vibrational control is achieved, (b) Shear Mode—in this mode the vibrational control is achieved by varying the shear force here one electrode is fixed and the other is free for rotation and (c) Squeeze Mode—in this mode the space between the electrodes is changed which presses the ER fluid results with a normal force.

2.1 Properties or electrorheological (ER) fluid

In electro rheological fluids there is a large reversible change in the colloidal suspension rheological properties when subjected to the external electric field. Lots of studies are present in which the principle and the uses of the electrorheological fluid are presented by many researchers across the globe. Another property of the ER fluids is that the response time of the ER fluid is very quick for the applied electric fields so the band width is thick. **Figure 1** represents the effect of ER fluid particles when application of electric field. For this interesting property the ER fluid has more demand is carious technological applications like smart structure, shock absorbers, engine mount and machine mount. The yield stress of the ER fluid can also be varied by introduction of the external electric field that is why it is also known as functional fluid. Winslow [2] patented the invention of the ER fluid. This ER effect is introduced in state of art automobile. The ER effect was first invented in 1942 by Winslow [2] after that the details understanding of the EF effect took lots of time and then to find the suitable solution for the ER fluid effect took further more time. The properties which delays and stops the ER fluid in few application fields are temperature stability, yield stress and power consumption. Particles size, carrier fluid properties, density, temperature and additives of the ER fluids plays a vital role for most of the properties changes of the ER fluids.

There is a limit up to which the dispersing particles can be mixed with the fluid because by increasing the concentration of the dispersing particles volume fraction the electrorheological effect of the solution increases which also causes few problems. As increasing the concentration of the dispersing particle after a certain concentration

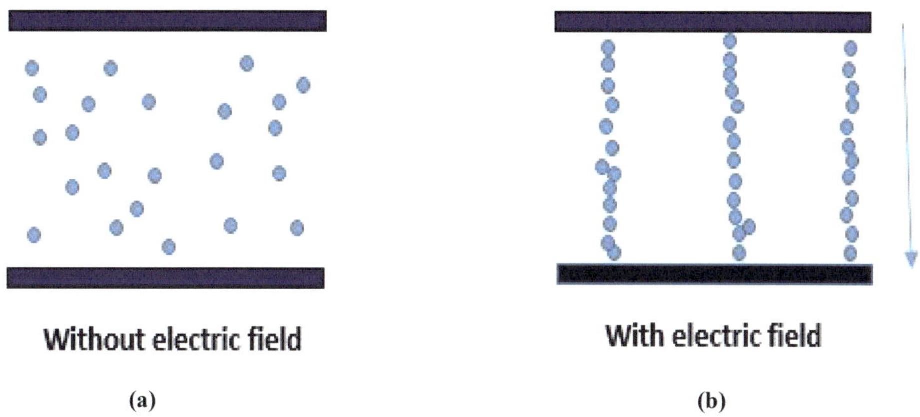

Figure 1.
(a) Dispersing particles without electric field, (b) dispersing particles with electric field.

limit the particles started settling down which cause a problem another problem which arises is the zero field viscosity increment. The viscosity is linked with the temperature i.e. the viscosity decreases when the temperature is increased. Temperature also decreases the dynamic yield strength. Mainly the change in the yield strength occurs due to relative permittivity and the conductivity of particle and also the chemical components of the fluid. Less amount of voltage approx. 1–4 KV/mm is needed for producing ER effect in the solution. 10^{-6} to 10^{-3} amp/cm^2 is the minimum needed current density for the ER effect. For calculating the power consumption of the suitable ER fluid the measurement of the current density are needed. Dynamic yield stress is one of the important ER fluid property, this stress is the maximum amount of stress required to flow the liquid when the electric field is applied. 100 Pa to 3 KPa is the range of the dynamic yield stress in current ER fluid. The comparison of the various ER fluids are still now difficult as because the standard testing procedure and the state for the fluid is not yet available properly and due to the dependency of the ER fluids on its dispersing particles and the fluid used combinations. For practical applications of the ER fluid the fluid must meet the desired criteria which are (a) Current density 4.0 KV/mm DC less than 10 μA/cm^2, (b) dynamic yield stress 4.0 KV/mm <3.0 KPa, (c) Zero field viscosity 0.1–0.3 Pas, i.e., 1–3 Poise, (d) Operational temp range −25°C to +125°C, (e) dielectric breakdown strength >50 KV/mm^2, (f) particle size 10 μm, (g) response time < millimeter, (h) Density 1–2 g/cm^3, (i) maximum energy density 0.001 Joule/cm^3, (j) power supply 2–5 KV@ 1–10 mA, (k) Any conductive surface material, (l) any opaque or transparent, and (m) physically and chemically stable with low conductivity and high breakdown voltage.

For shear loading state applications usually the ER materials are used. The relationship between the ER material and the share are shown in the **Figure 2**. In the year 1949 Winslow [2] invented the post-yield appearance of the ER effect. During that time the materials which behave like changing in viscosity were called electro-viscous fluids as their effective or actual viscosity changes were noticeable macroscopically. Many years after it was investigated that with the change in the applied electric field the apparent or the effective viscosity υ remains constant, only the noticeable change was found out was the yield stress of the Bingham plastic suspension. This is shown in **Figure 2**. Ideal plastic fluids are also another name given to the Bingham plastics, i.e., this fluid does not have viscosity (zero viscosity). A formula representing the shear stress exceeds the yield stress of the material is given by $\tau = \tau y + \vartheta\gamma$, where τ represents Shear stress, τy represents Yield Stress and $\vartheta\gamma$ represents Shear Strain. The behavior of the ER material the comparison of the

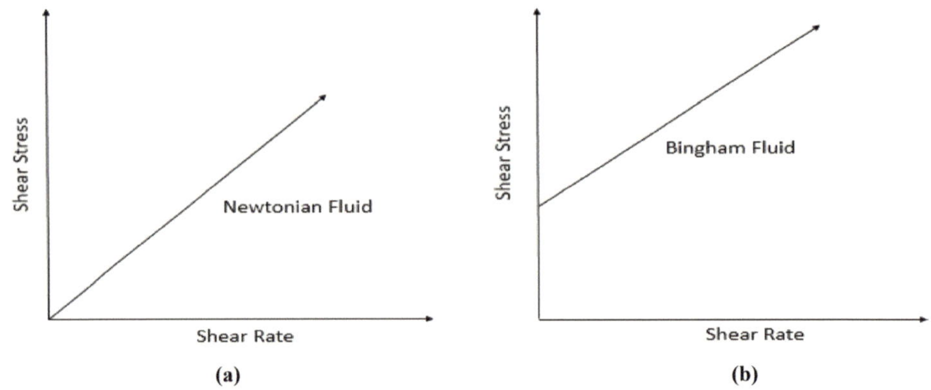

Figure 2.
Smart fluid characterization (a) without electric field and (b) with electric field.

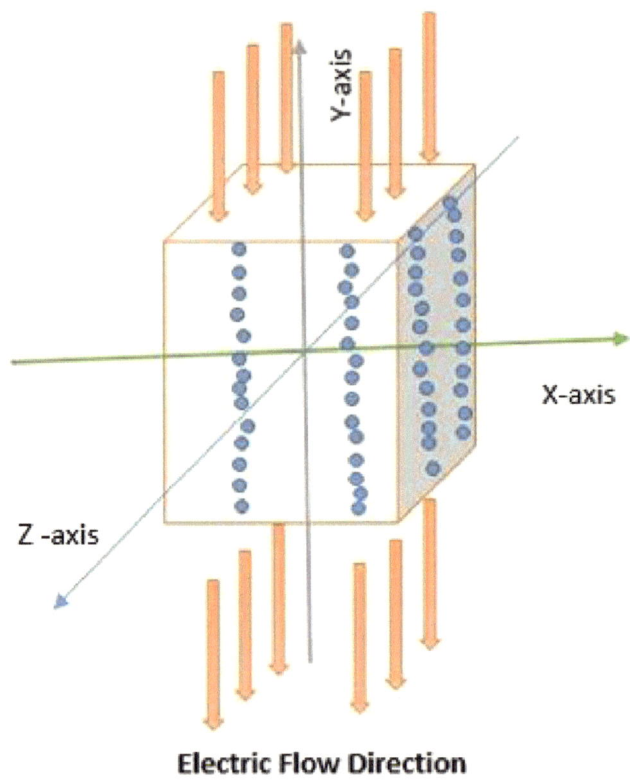

Figure 3.
Reaction of the ER fluid when external electric field is applied.

post yield behavior still not investigated. With increasing in the electric field the shear yield stress increases while the yield strain remains 1% for almost all fields. The reaction of the ER fluid on electric field is shown in **Figure 3**.

2.2 Typical carrier fluid and particles

The ER fluids which are available in the markets are very costly so here are few lists of combinations of the additive particles with the fluid to prepare the cost effective ER fluid. With suitable proportions and amount of the additive particle we can achieve the desired ER fluid as per our need. Various carrier fluids are aldehyde, grease, ketones, kerosene, aroclor, castor oil, chloroform, mineral oil, olefins, olive oil, dielectric oil, diphenyl sebacate, various ethers, resin oil, transformer oil, silicon oil etc. Various additive particles for the ER fluid are alfa silica, alginic acid, alumina, alfa methylacrylate, mannitol, boron, macrocel-C, carbon, cellulose, charcoal, chlorides, dyes, gypsum, micronized mica, nylon powder, olefins, porhin, pyrogenic silica, quartz, rubber, silica gel, etc. [3].

2.3 Preparation of ER fluids

ER fluid preparation procedure are very simple and mostly all the ER fluids are prepared by this manner the following procedure is used for preparing the ER fluids: (a) The desired powder is chosen and same particle powder size particles are required for the ER fluid dispensing particle, (b) the chosen powder must be passed through size sieve for all the particles same and must be weighted on the weighing machine, (c) the powder is poured in glass container and desired amount of the ER fluid is poured in the glass container which contains the powder of uniform size and are stirred continuously until the powder mixed with the fluid completely, (d) the mixture of the powder and the fluid are stirred for 2 h by glass rod or magnetic starrer at a constant RPM to get a uniform homogenous mixture, (e) the mixed solution is passed to a vane pump five times to get a good result homogenous solution and (f) this process should be followed for other ER solution preparation [3].

2.4 Testing and selection of ER fluids

The testing of the ER fluid is necessary for selecting of the desired ER fluid for the desired application. The following tests are mainly used (a) Temperature test, (b) breakdown test, (c) viscosity test and (d) sedimentation test.

2.5 Applications of electro rheological fluids

The electrorheological fluids which are totally dependent on the applied electric fields are used in resistive force creation and damping. Examples of applications are active vibration suppression and motion control. Wang et al. [4] have presented the uses of ER fluids in microfluidics [5]. Various industries like automobiles industries are demanding modified ER fluids with more efficiency Gurka et al. [6] introduced ER-Fluid RheOil®3.0 which improves the sedimentation and re-dispersing behavior. Brennan et al. [7] studied and distinguished the two classes of the ER dampers, first one acts by shearing the stationary fluid and the second one acts by pumping the ER fluid [5]. The two classes are described in details below. Most of the dampers of smart fluids have three common components, i.e., a cylinder, cylinder valve housing and a piston. The vibrating structure kinetic energy can be controlled and dissipated by providing either electric or magnetic field in the valve. In the ER damping process two types of frictions are used they are viscous and coulomb friction [8]. The columbic force denotes the friction acting when two surfaces comes in contact to each other like friction of bearing and hinges friction. Friction is independent to the body velocity, i.e., it is constant. To push fluids through narrow obstructive passage viscous friction comes into play these exists in valves and orifices and is body velocity dependent. The viscous friction and the columbic friction summation is the actuation friction which is denoted in **Figure 4**. These frictions have good effects also in the damping machines. The transmission of the vibration to the device is possible by dry sealing friction. For sensitive instruments small vibrations can cause poor accuracy [9]. Bad effect of the friction is also present in the system when the force applied is near to overwhelm the static friction this is known as motion of stick–slip.

At a near to zero velocity the stick–slip motion happens like an unexpected motion of jerking. Naturally, kinetic friction coefficient in between the two surfaces is smaller than the static friction coefficient. When the given force is more to overwhelm than the static friction then the friction decreases from static to dynamic. Because of this sudden decrease of the friction there will be a sudden velocity jump movement. To show this effect the system of two degree of freedom is taken.

2.5.1 Shear mode

In this type of mode of ER damper there are one or two parallel electrodes which can move parallel to each other and is always perpendicular to the electric field applied so that the fluid can have uniform shear and the ER fluid is present in between the two electrodes. From **Figure 5a** c and l are the breath and length of the electrode and j is the electrodes gap. Here E is given voltage, F is net damping force and V is the relative velocity of the electrodes. Two forces are acting in this ER damper (a) Active force Fc because of ER effect and (b) Passive force Fy due to the fluid viscosity. Fy, i.e., the passive force is always present and directly linked with the viscosity of the fluid as well as the damper geometric properties. During application of the electric field a force Fc (because of creation of particles suspension lining up between the electrodes) i.e. static force which is needed to overwhelmed so that the motion can occur [10]. The force Fc is product of area of electrode and the yield strength of the fluid and does not depend on the electrode plate velocity. The net force F of damping of this ER damper is the sum of two components of force. The main aim of this ER damper is to give large ratio of off-field to on-field damping by force ratios Fy and Fc. Because of this large ratio gives various responses by ER unit with changing voltage.

2.5.2 Valve mode

In this type of mode the ER fluid is pressed between the two electrodes as given in **Figure 5b**. Because of this the ER fluid is exposed to tensile, compression as wells as shear. In the absence of the given electric field if the ER fluid is pressed it behaves like Newtonian fluid. There is a pressure drop AP occurs at flow rate volume Q. This pressure change in between the valve is because of the velocity of the ER fluid. Moreover, during the presence of the electric field, yield stress is generated by the ER fluid which results more pressure drop between the electrodes plates length. The net damping force is summation of two force components of this type of ER

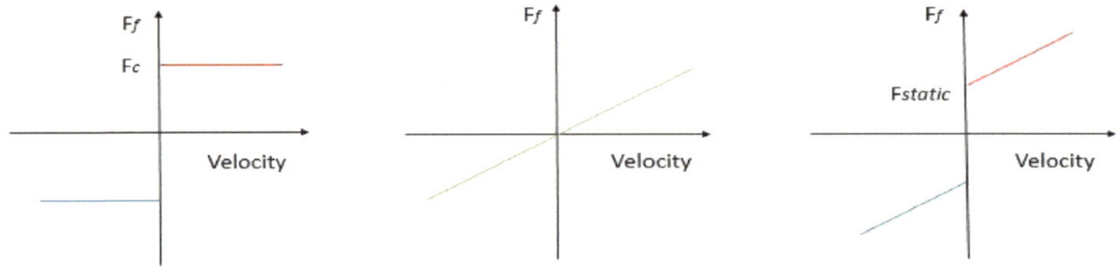

Figure 4.
Actuator friction (a) friction columbic, (b) friction viscous, and (c) total friction.

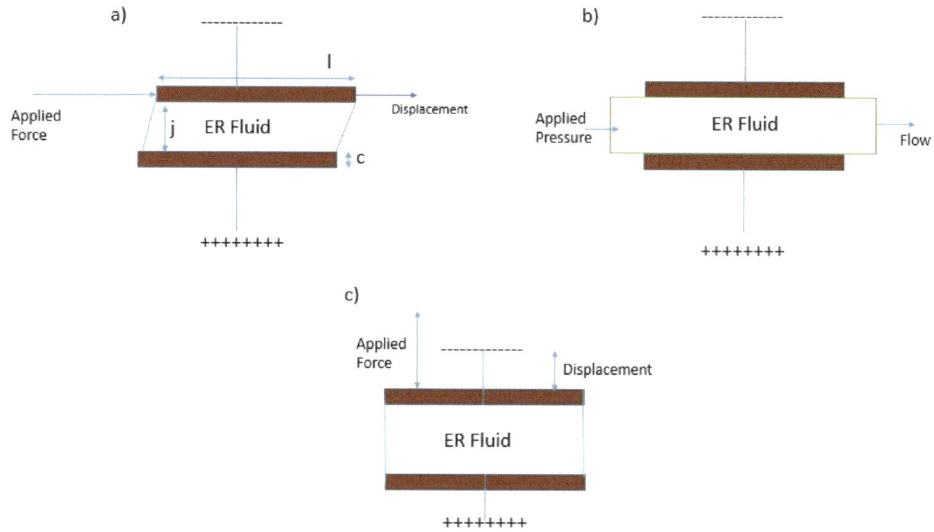

Figure 5.
Modes of operation: (a) shear, (b) valve, and (c) squeeze.

damper. In this type of mode the device effectiveness is the across valves pressure drop with or without the effect of ER [10].

The electrorheological fluids which are totally dependent on the applied electric fields are used in resistive force creation and damping. Examples of applications are active vibration suppression and motion control. L. Wang et al. [4] have presented the uses of ER fluids in microfluidics. Various industries like automobiles industries are demanding modified ER fluids with more efficiency Gurka et.al [6] introduced ER-Fluid RheOil®3.0 which improves the sedimentation and re-dispersing behavior. Brennan et al. [7] studied and distinguished the two classes of the ER dampers, first one acts by shearing the stationary fluid and the second one acts by pumping the ER fluid. The two classes are described in details below.

2.5.3 Squeeze mode

In this mode the gap between the electrodes are changed and the ER fluid is pressed or squeezed by the force acting normally. **Figure 5c** represents the squeezing mode of the ER fluid.

2.5.4 Applications of ER fluid technique in engineering field

ER fluids have wide applicability, economic benefit, social benefit high performance for these advantages these smart fluids will find path in various engineering applications in various technological fields. Without any doubt we can say in the future ER technology is going to rule various applications in engineering technological fields. As soon as this technology is accepted then it will be a revolution in both economy and society. From all these advantages of the ER fluids we can predict that in the near future the ER fluids will be used in various technological fields as given below.

2.5.4.1 Automobile/motor vehicle industry

Scientists and Engineers can develop new kind of parts that can easily fulfill the needs of the motor vehicles using the technology of ER. Like for example ER technology used for cooling engine i.e. speed fan clutch of the motor vehicle, shock absorber, brake having break torque controlled, system for suspensions by damping controlled etc., These components using ER technologies will have less wear and tear, more performance, less cost, prolong life service, controlled easily, easy to produce by microcomputer, fast response, high sensitivity.

2.5.4.2 Hydraulic industry

The valves which are used nowadays for control of pressure and flow rate control can be replaced by ER technology in the future. Because ER technology valves will have no or less movable parts, simple easy structure, low cost, prolong service life, no mechanical processing, minimal tear and wear and electronical control of pressure and rate of flow. For this reasons ER technology will rule the hydraulic industry in the near future.

2.5.4.3 Fluid sealing field

By utilizing the benefits of the ER technology engineers can produce new type of rotational sealing controlled devices for face the challenges of the magnetic fluid sealing and rubber fluid sealing. Because of the pros like good effect of sealing, minimal tear and wear, less magnetic field and prolong life of service.

2.5.4.4 Robotic industry

In robotic industries nowadays for flexible joints are being controlled by hydro-electric control devises instead of ER fluidic joints technology which can perform much better function than the hydraulic-electric control. Engineers are designing and manufacture flexible joints which will have less volume, fast response time, minimal wear as well as tear, nimble, and which can be easily controlled by microcomputers. ER fluids can provide all these advantages over the hydraulic-electric controls.

2.5.4.5 Commercial uses

There are various commercial uses of the ER fluids and many uses are still undiscovered, in automotive industries the ER fluids are used in clutches, seat dampers, shock absorber, engine mount etc. Many other applications of the ER fluids are listed as follows: (a) Fluid flow via thin channel, (b) for friction instruments clutches, (c) servomechanism for impact and vibrator instruments, (d) pick-pick applications, (e) damping isolator, (f) automobile damping, (g) mounts for engine, (h) power transmission in robots, (i) machine tool artificial intelligence, etc. This list is not the final list because still now many uses of the ER fluid in various fields are yet to discover.

3. Characterization of the ER fluid

3.1 Rheological characterization

Rheological characterization is done to identify the change in viscosity of the ER fluid with respect to the shear rate at various electric fields. Garcia et al. [11]

have studied the rheological properties of the ER fluid by using ARES rheometer by using parallel plate diameter 50 mm diameter electrode with 1 mm gap between them. High voltage amplifier was used to supply the DC voltage.

3.2 Dielectric characterization

To study the permittivity and the power factor of the ER fluid the dielectric properties characterization are done. Rejon et al. [12] describes the method of measuring the dielectric properties of the ER fluid. They used guard ring capacitors and high resistor meter. DC high voltages were used for the test.

3.3 Structural characterization

The structural changes of the ER field during and before the DC voltage was studied by Rejon et al. [13]. The studied the microscopic structure of the ER fluid by microscope. They studied the microstructural changes of the ER fluid at different DC applied voltages from 0.5 to 2.5 KV/mm.

4. Conclusion

ER fluids have lots of interesting properties which attracts them in various applications fields among the various important properties of the ER fluid lies fast reaction, precise controllability and easy boundary between the electrical and mechanical input output power. Because of these interesting properties of the ER fluid the ER fluid is used in motion control and will be used in various applications fields in the near years to come. ER fluids characteristics in most advanced way is briefly described below as given in latest reports: (a) When external electric field is given ER effect is seen by change in viscosity of the carrier fluid from liquid to solid as the viscosity of the liquid increases and after removal of the electric field solid to liquid viscosity decreases making the liquid less thick like the initial state, (b) the process in which the ER fluid changes its state from liquid to solid upon application of the electric field must be reversible, i.e., it should return back to its original state (liquid state) as soon as the external electric field is removed. Viscosity change must be less step, (c) upon application of the electric field the transition of the liquid state to the solid state must be very fast, i.e., 5–10 s, (d) and liquid to solid transition must be only possible by electric field only and not by any other means. By all these characteristics of the ER fluid the ER fluid can be connected with the modern technological applications. This technology is one newly type of future challenge as its attractive properties are being used broadly, which can bring a big change in industries. The main component of the ER technology is the ER fluid which should bring in the technological applications like dampers of ER fluids which is a best solution for control of vibrations.

Author details

Sudipto Datta[1*], Ranjit Barua[1] and Jonali Das[2]

1 Centre for Healthcare Science and Technology, Indian Institute of Engineering Science and Technology, Shibpur, Howrah, WB, India

2 Raja Peary Mohan Collage, Department of Chemistry, Calcutta University, Uttarpara, Hooghly, WB, India

*Address all correspondence to: dattadip440v@gmail.com

References

[1] Ko YG, Choi US. Negative electrorheological fluids. 2013;**1655**

[2] Wilson WM. Induced vibration of suspensions using ER. Journal of Applied Physics. 1947;**20**:1137-1140

[3] Jadhav A, Ashta T. ER. 2017

[4] Wang L, Gong X, Wen W. Electrorheological fluid and its applications in microfluidics. 2011:91-115

[5] Elderrat HI. Research towards the design of a novel SMART fluid damper using a McKibben actuator. 2013

[6] Yamanaka M et al. New electrorheological fluids—Characteristics and implementation in industrial and mobile applications. 2009

[7] An electrorheological fluid vibration damper;**83**

[8] Yang G. Large-scale MR fluid dampers: Modeling and dynamic performance considerations. 2002;**24**:309-323

[9] Seong MS et al. Design and performance evaluation of MR damper for integrated isolation mount. 2015;**22**

[10] Khanicheh A, Mintzopoulos D, Weinberg B, Tzika AA. Evaluation of electrorheological fluid dampers for applications at 3-T MRI environment. 2008

[11] Rejon-Garcia L. Rheological and dielectric characterization of electrorheological fluids composed of both dispersed particles and liquid drops in a dielectric medium. 2015:2002

[12] Rejon L, Castaneda-Aranda I, Manero O. Colloids and Surfaces A: Physicochemical and Engineering Aspects. 2001;**182**:93-107

[13] Rejon L, Ponce MA, De la Luz C, Nava R, Castano VM. Journal of Materials Science: Materials in Electronics. 1996;**7**:433-436